AMERICAN MATHEMATICAL SOCIETY
TRANSLATIONS

Series 2

Volume 119

Twelve Papers in Algebra

by

B. M. Šaĭn

Hsieh Sheng-kang

V. A. Dem′janenko

A. A. Šmelev

Wang Yuan

I. Š. Slavutskiĭ

S. A. Ovsienko

A. V. Roĭter

A. I. Mal′cev

A. Ju. Ol′šanskiĭ

B. I. Plotkin

A. I. Širšov

AMERICAN MATHEMATICAL SOCIETY
PROVIDENCE, RHODE ISLAND
1983

Edited by LEV J. LEIFMAN

1980 *Mathematics Subject Classifications*. Primary 06A12, 08A30, 10B15, 10F37, 10G20, 10J15, 12A25, 12A35, 12A50, 12A70, 12B30, 14H45, 15A63, 16A64, 16A68, 16A90, 17C50, 17D05, 20E10, 20E26, 20E34, 20M99; Secondary 06A15, 16A22, 17D05, 20D99, 20F16, 20M14.

Library of Congress Cataloging in Publication Data

Main entry under title:
Twelve papers in algebra.

(American Mathematical Society translations; ser. 2, v. 119)

1. Algebra—Addresses, essays, lectures. I. Šaĭn, B. M. II. Leĭfman, L. IA. (Lev IAkovlevich) III. Series.

QA3.A572 Ser. 2, vol. 119	510s	82-24434
[QA155.2]	[512]	

ISBN 0-8218-3074-0
ISSN 0065-9290

TABLE OF CONTENTS

RUSSIAN AND CHINESE TABLE OF CONTENTS[*]

---

[*]The American Mathematical Society scheme for transliteration of Cyrillic may be found at the end of index issues of *Mathematical Reviews*.

Amer. Math. Soc. Transl.
(2) Vol. **119**, 1983

Pseudosemilattices and Pseudolattices*

B. M. ŠAĬN [BORIS M. SCHEIN]

The connection between commutative idempotent semigroups and the theory of ordered sets, where every two elements have a greatest lower [or least upper] bound, is well known. The basic idea of this paper is an extension of this connection to a wider class of idempotent semigroups, the so-called *normal idempotent semigroups* (or *normal bands*), i.e. to idempotent semigroups satisfying the identity $xyzx = xzyx$. As an analog of ordered sets one considers *double quasi-ordered sets*, i.e. systems of the form $(A; \zeta_1, \zeta_2)$, where ζ_1 and ζ_2 are two quasi-order relations on a set A. An element a is called a *minorant* of an ordered pair (a_1, a_2) of elements of A if $a \leqslant a_1(\zeta_1)$ and $a \leqslant a_2(\zeta_2)$. A minorant of (a_1, a_2) which is greater (relative to both ζ_1 and ζ_2) than all other minorants is called the *greatest lower bound* (g.l.b.) of (a_1, a_2). If $\zeta_1 \cap \zeta_2$ is an order relation, then each ordered pair has at most one g.l.b. If the g.l.b. exists for every ordered pair of elements of A, then A endowed with the operation of forming the g.l.b. forms an algebra, called a *lower pseudosemilattice*. Quite similarly one defines the operation of forming the *least upper bound* (l.u.b.) and *upper pseudosemilattices*. It turns out that the ordinary semilattices can be characterized in the class of all pseudosemilattices by commutativity of their operation. If one considers both operations of forming the g.l.b. and l.u.b. together, one arrives at algebras which we call *pseudolattices*.

Within the class of all pseudosemilattices we pick out the so-called *proper pseudosemilattices* and also a narrower subclass of *associative pseudosemilattices*. Such systems are characterized by systems of identities. The structure of associative pseudosemilattices and pseudolattices is completely described.

Note that the problem of generalization of lattice theory for the noncommutative case has drawn more and more attention recently, dozens of papers have considered it (e.g., one may consult [1] where further references can be found; see

---

* Translation of Izv. Vysš. Učebn. Zaved. Matematika **1972**, no. 2(117), 81–94; MR **46** #5203.
1980 *Mathematics Subject Classification*. Primary 06A12, 20M99.

[2] for applications of noncommutative lattices). Partly such interest can be explained by P. Jordan's idea of using idempotent semigroups and noncommutative lattices as a mathematical apparatus of quantum mechanics. Repeatedly (see [1]) the desirability of establishing a connection between the theory of noncommutative lattices and that of ordered sets in analogy with the theory of ordinary lattices has been mentioned, but so far no satisfactory connection has been established. A special class of noncommutative lattices has been connected with double quasi-ordered sets [3]; however, this connection seems to be much more artificial than that established in the present paper. As far as we know the first results on the connection of the theory of noncommutative pseudosemilattices with that of double quasi-ordered sets are contained in [4], where some results of the present paper were announced.

A double quasi-ordered set $(A; \zeta_1, \zeta_2)$ is called *strict* if $\zeta = \zeta_1 \cap \zeta_2$ is antisymmetric, i.e. an order relation. An element $a \in A$ is called a *majorant* of a pair (a_1, a_2) if $a_1 \leqslant a(\zeta_1)$ and $a_2 \leqslant a(\zeta_2)$. An element a is called the *least upper bound* of (a_1, a_2) if a is a majorant of this pair and, for every majorant a_0 of that pair, $a < a_0(\zeta)$. Double quasi-ordered sets $(A; \overset{-1}{\zeta}_1, \overset{-1}{\zeta}_2)$ and $(A; \zeta_2, \zeta_1)$ are called *dual* and *inverted* relative to $(A; \zeta_1, \zeta_2)$ respectively. It is easily seen that *the following three properties are equivalent for any double quasi-ordered set* $(A; \zeta_1, \zeta_2)$:

(1) *A is strict.*

(2) *Every pair of elements of A possesses at most one g.l.b. in A.*

(3) *Every pair of elements of A possesses at most one l.u.b. in A.*

In what follows we suppose that all double quasi-ordered sets considered are strict; we call them SDQS, for short. The g.l.b. of a pair (a_1, a_2), if it exists, we denote by $a_1 \wedge a_2$, and the l.u.b. we denote by $a_1 \vee a_2$. Then $\wedge$ and $\vee$ are partial binary operations on A. One can easily verify that both these operations are *idempotent*, i.e. $a \wedge a$ and $a \vee a$ always exist and equal a.

If $\zeta_1 = \zeta_2$, one can identify $(A; \zeta_1, \zeta_2)$ and $(A; \zeta)$; then the SDQS becomes an ordinary ordered set and g.l.b.'s and l.u.b.'s have their usual meaning. In this case ordinary lower and upper semilattices correspond to lower and upper pseudosemilattices, and lattices correspond to pseudolattices.

A twofold duality principle holds for SDQS's. It consists in transition to dual or inverted SDQS's respectively. When one considers the dual SDQS, the operations $\wedge$ and $\vee$ interchange their roles, and if one considers the inverted SDQS, then $a_1 \wedge a_2$ and $a_1 \vee a_2$ are the g.l.b. and l.u.b. of the inverted pair (a_2, a_1) rather than of the pair (a_1, a_2). We will use these duality principles quite often, without specially mentioning them.

EXAMPLE 1. Let $A = \{0, 1, 2, 3, 4\}$ and consider the binary relations

$$\zeta_1 = (\{0, 3\} \times \{0, 2, 3, 4\}) \cup (\{1\} \times \{1, 2, 4\}) \cup (\{2, 4\} \times \{2, 4\});$$

$$\zeta_2 = (\{0, 1, 2\} \times A) \cup (\{3, 4\} \times \{3, 4\}); \qquad \zeta = \zeta_1 \cap \zeta_2;$$

$$\bar{\zeta}_1 = (\{0, 3\} \times \{0, 3\}) \cup (\{1\} \times \{1, 2, 4\}) \cup (\{2, 4\} \times \{2, 4\}); \quad \bar{\zeta} = \bar{\zeta}_1 \cap \zeta_2.$$

One can easily see that $(A; \zeta_1, \zeta_2)$ and $(A; \bar{\zeta}_1, \zeta_2)$ are SDQS's, and a simple computation shows that the same pseudolattice $(A; \wedge, \vee)$ whose operations $\wedge$ and $\vee$ are given by the Cayley tables

$$
\begin{array}{l}
0\,0\,0\,3\,3 \\
1\,1\,1\,1\,1 \\
2\,2\,2\,4\,4 \\
0\,0\,0\,3\,3 \\
2\,2\,2\,4\,4
\end{array}
\qquad\qquad
\begin{array}{l}
0\,0\,0\,3\,3 \\
1\,1\,1\,4\,4 \\
2\,2\,2\,4\,4 \\
0\,0\,0\,3\,3 \\
2\,2\,2\,4\,4
\end{array}
$$

(here both rows and columns follow the usual order, so we have omitted their numeration) corresponds to each of these SDQS's.

EXAMPLE 2. Let $A = \{0, 1, 2, 3\}$ and define two order relations ζ_1 and ζ_2 respectively by the relations $0 < 1 < 2$ and $0 < 3$ relative to ζ_1 and $0 < 1 < 2$ and $1 < 3$ relative to ζ_2. Then $\zeta_1 \subset \zeta_2$ and $\zeta = \zeta_1 \cap \zeta_2 = \zeta_1$. The lower pseudosemilattice $(A; \wedge)$ whose operation is given by the Cayley table

$$
\begin{array}{l}
0\,0\,0\,0 \\
0\,1\,1\,1 \\
0\,1\,2\,1 \\
0\,0\,0\,3
\end{array}
$$

is associated with $(A; \zeta_1, \zeta_2)$.

Now suppose that $\mathfrak{A} = (A; \cdot)$ is an algebra with a binary operation $\cdot$, and define the binary relations $\zeta_1^{\cdot} = \{(a_1, a_2) \mid a_2 a_1 = a_1\}$ and $\zeta_2^{\cdot} = \{(a_1, a_2) \mid a_1 a_2 = a_1\}$ on A. The relative $(A; \zeta_1^{\cdot}, \zeta_2^{\cdot})$ is called the *absorption relative* of $\mathfrak{A}$. Clearly, if the operation $\cdot$ is idempotent, then the relations $\zeta_1^{\cdot}$ and $\zeta_2^{\cdot}$ are reflexive. However, these relations need not be transitive even for a pseudosemilattice, as Example 1 shows, for then $(0, 1), (1, 3) \in \zeta_2^{\wedge}$, but $(0, 3) \notin \zeta_2^{\wedge}$; and also $(2, 4), (4, 1) \in \zeta_1^{\vee}$, but $(2, 1) \notin \zeta_1^{\vee}$. At the same time, if a lower pseudosemilattice $(A; \cdot)$ is associated with an SDQS $(A; \zeta_1, \zeta_2)$, then, as one can easily see, $\zeta_1^{\cdot} \subset \zeta_1$ and $\zeta_2^{\cdot} \subset \zeta_2$.

An algebraic system is called a *pseudosemilattice* if it is isomorphic to an upper (and so also to a lower) pseudosemilattice.

PROPOSITION 1. *Suppose that $(A; \cdot)$ is a pseudosemilattice and $(A; \zeta_1^{\cdot}, \zeta_2^{\cdot})$ is its absorption relative. The relation $\zeta_1^{\cdot}$ is transitive (i.e. it is a quasi-order) if and only if $x((xy)z) = (xy)z$. The relation $\zeta_2^{\cdot}$ is transitive if and only if $(x(yz))z = x(yz)$.*

PROOF. First we prove that every pseudosemilattice satisfies the identities

$$x(xy) = xy, \qquad (xy)y = xy. \tag{I}$$

Indeed, let $(A; \cdot)$ be the pseudosemilattice associated with an SDQS $(A; \zeta_1, \zeta_2)$. Then $xy \leqslant x(\zeta_1)$ and $xy \leqslant xy(\zeta_2)$; hence $xy \leqslant x(xy)(\zeta)$. On the other hand, $x(xy) \leqslant x(\zeta_1)$ and $x(xy) \leqslant xy(\zeta_2)$, and also $xy \leqslant y(\zeta_2)$, so that $x(xy) \leqslant y(\zeta_2)$, i.e. $x(xy) \leqslant xy(\zeta)$. It follows from the antisymmetricity of ζ that $x(xy) = xy$. The second identity follows dually.

The transitivity of $\zeta_1^{\cdot}$ means that $(a, b), (b, c) \in \zeta_1^{\cdot} \to (a, c) \in \zeta_1^{\cdot}$, or, what is the same, $ba = a, cb = b \to ca = a$. We set $a = ((xy)z), b = xy$ and $c = x$. Then $ba = a$ and $cb = b$ by (I). If $\zeta_1^{\cdot}$ is transitive, then $ca = a$, i.e. the identity of the proposition holds. But if $x((xy)z) = (xy)z$, then $ba = a$ and $cb = b$ imply $ca = c(ba) = c((cb)a) = (cb)a = ba = a$. The second assertion of Proposition 1 is dual to the first.

A pseudosemilattice is called *proper* if it satisfies the identities of Proposition 1. Defining $\zeta^{\cdot} = \zeta_1^{\cdot} \cap \zeta_2^{\cdot}$, we see that if $(A; \cdot)$ is proper, then $\zeta^{\cdot}$ is always antisymmetric, for $(a, b) \in \zeta^{\cdot}$ and $(b, a) \in \zeta^{\cdot}$ mean that $a = ba = b$. Therefore an SDQS $(A; \zeta_1^{\cdot}, \zeta_2^{\cdot})$ is naturally associated with each proper pseudosemilattice $(A; \cdot)$.

THEOREM 1. *A lower pseudosemilattice is associated with the SDQS which is its absorption relative if and only if it is proper.*

PROOF. If the absorption relative of a pseudosemilattice is an SDQS, then, by Proposition 1, the pseudosemilattice is proper. To prove the second part of our theorem we need the following two lemmas.

LEMMA 1. *Every pseudosemilattice satisfies the identities $(xy)z = (xy)(xz)$ and $x(yz) = (xz)(yz)$.*

PROOF. Let $(A; \cdot)$ be the lower pseudosemilattice associated with an SDQS $(A; \zeta_1, \zeta_2)$. Then $(xy)z \leqslant xy(\zeta_1)$ and $xy \leqslant x(\zeta_1)$, i.e. $(xy)z \leqslant x(\zeta_1)$. Moreover, $(xy)z \leqslant z(\zeta_2)$; hence $(xy)z \leqslant xz(\zeta)$, and so $(xy)z \leqslant xz(\zeta_2)$. Since $(xy)z \leqslant xy(\zeta_1)$, we obtain $(xy)z \leqslant (xy)(xz)(\zeta)$. On the other hand, $(xy)(xz) \leqslant xy(\zeta_1)$ and $(xy)(xz) \leqslant xz(\zeta_2)$, $xz \leqslant z(\zeta_2)$; hence $(xy)(xz) \leqslant z(\zeta_2)$ and $(xy)(xz) \leqslant (xy)z(\zeta)$. Therefore $(xy)z = (xy)(xz)$. The second identity is dual to the first.

LEMMA 2. *Every pseudosemilattice $(A; \cdot)$ has the property $\xi_1^{\cdot}\langle x\rangle \cap \xi_2^{\cdot}\langle y\rangle \subset \xi^{\cdot} \langle xy\rangle$, where $\xi_1^{\cdot} = \zeta_1^{-1}$, $\xi_2^{\cdot} = \zeta_2^{-1}$, $\xi^{\cdot} = \zeta_1^{-1}$ and $\rho\langle x\rangle = \{y \mid (x, y) \in \rho\}$.*

PROOF. Let $z \in \xi_1^{\cdot}\langle x\rangle \cap \xi_2^{\cdot}\langle y\rangle$, i.e. $(z, x) \in \zeta_1^{\cdot}$ and $(z, y) \in \zeta_2^{\cdot}$. In other words, $z = xz$ and $z = zy$. By Lemma 1, $(xy)z = (xy)(zy) = x(zy) = xz = z$ and $z(xy) = (xz)(xy) = (xz)y = zy = z$, i.e. $z \in \xi^{\cdot}\langle xy\rangle$.

Turning now to the proof of Theorem 1, suppose that $(A; \cdot)$ is a proper lower pseudosemilattice. By Proposition 1, its absorption relative is an SDQS. Let $z \in \xi^{\cdot}\langle xy\rangle$, i.e. $(z, xy) \in \zeta^{\cdot}$; that is, $z(xy) = z$ and $(xy)z = z$. It follows that $z = (xy)z = x((xy)z) = xz$ and $z = z(xy) = (z(xy))y = zy$, i.e. $(z, x) \in \zeta_1^{\cdot}$ and $(z, y) \in \zeta_2^{\cdot}$; that is, $z \in \xi_1^{\cdot}\langle x\rangle \cap \xi_2^{\cdot}\langle y\rangle$. By Lemma 2, $\xi^{\cdot}\langle xy\rangle = \xi_1^{\cdot}\langle x\rangle \cap \xi_2^{\cdot}\langle y\rangle$, i.e. the set of all minorants of the ordered pair (x, y) in $(A; \zeta_1^{\cdot}, \zeta_2^{\cdot})$ coincices with the set of all elements which are less that xy relative to $\zeta^{\cdot}$, i.e. xy is the g.l.b. of (x, y). Q.E.D.

Unfortunately, one and the same pseudosemilattice or pseudolattice can be associated with different SDQS's, as Example 1 shows. In this respect proper pseudosemilattices exhibit the closest analogy with ordinary semilattices, which are also associated with their absorption relatives.

Pseudosemilattices and pseudolattices need not be associative. In Example 1, $(1 \wedge 3) \wedge 1 = 4 \wedge 1 = 2$, but $1 \wedge (3 \wedge 1) = 1 \wedge 0 = 1$; and $(3 \vee 1) \vee 3 = 0 \vee 3 = 3$, but $3 \vee (1 \vee 3) = 3 \vee 1 = 0$, i.e. even the identity $(xy)x = x(yx)$, which is weaker than associativity, does not hold either for $\wedge$ or for $\vee$. Both operations $\wedge$ and $\vee$ are anticommutative in Example 1, i.e. $x \wedge y = y \wedge x$ only if $x = y$, and $x \vee y = y \vee x$ only if $x = y$. One can easily check that the pseudosemilattice given in Example 2 is proper, for it satisfies the condition of Proposition 1. However, it is not associative, because $(1 \wedge 3) \wedge 1 = 1 \wedge 1 = 1$ and $1 \wedge (3 \wedge 1) = 1 \wedge 0 = 0$. In this case both $\zeta_1^\cdot$ and $\zeta_2^\cdot$ are order relations.

It is noteworthy that if $((A; \zeta_1^i, \zeta_2^i))_{i \in I}$ is a family of SDQS's with which the same pseudosemilattice is associated, then, as is easily seen, this pseudosemilattice is associated with the SDQS $(A; \bigcap_{i \in I} \zeta_1^i, \bigcap_{i \in I} \zeta_2^i)$ as well.

The following is an abstract characterization of proper pseudosemilattices.

THEOREM 2. *An algebra with a binary operation is a proper pseudosemilattice if and only if it it satisfies the identities* $xx = x$, $(xy)(xz) = (xy)z$, $(xz)(yz) = x(yz)$, $x((xy)z) = (xy)z$ *and* $(x(yz))z = x(yz)$.

PROOF. A proper pseudosemilattice satisfies these identities by Proposition 1 and Lemma 1. Now let an algebra $(A; \cdot)$ satisfy these identities. Then $x(xy) = (xx)(xy) = (xx)y = xy$. Analogously we check the second identity in (I). Using (I) and the idempotence, exactly as in the proof of Proposition 1 we show that $(A; \zeta_1^\cdot, \zeta_2^\cdot)$ is an SDQS. Now we repeat without any difficulty the proof of Lemma 2 for $(A; \cdot)$. Then, for $(A; \zeta_1^\cdot, \zeta_2^\cdot)$, we repeat the second part of the proof of Theorem 1. As a result, $(A; \cdot)$ turns out to be the lower pseudosemilattice associated with $(A; \zeta_1^\cdot, \zeta_2^\cdot)$. Theorem 2 is proved.

The problem of abstract characterization of arbitrary pseudosemilattices remains open. Besides the properties of arbitrary pseudosemilattices which we have already mentioned, we may point out that, $(A; \cdot)$ being a pseudosemilattice, the relations $\zeta_1^\cdot \cap \xi_1^\cdot$ and $\zeta_2^\cdot \cap \xi_2^\cdot$ are transitive, i.e. they are equivalence relations. Indeed, let $(A; \cdot)$ be a lower pseudosemilattice associated with an SDQS $(A; \zeta_1, \zeta_2)$, and let $\eta_l = \{(a_1, a_2) \mid a_1 a = a_2 a$ for all $a \in A\}$ and $\eta_r = \{(a_1, a_2) \mid aa_1 = aa_2$ for all $a \in A\}$ (the relations η_l and η_r are called relations of *left* and *right interchangeability*, respectively, on $(A; \cdot)$). Supposing that $\xi_1 = \zeta_1^{-1}$ and $\xi_2 = \zeta_2^{-1}$, we have $\eta_l = \zeta_1^\cdot \cap \xi_1^\cdot = \zeta_1 \cap \xi_1$ and $\eta_r = \zeta_2^\cdot \cap \xi_2^\cdot = \zeta_2 \cap \xi_2$. Indeed, if $(a_1, a_2) \in \eta_l$, then $a_1 a = a_2 a$ for all $a \in A$. If $a = a_1$, we have $a_1 = a_2 a_1$, i.e. $(a_1, a_2) \in \zeta_1^\cdot$. It follows from the obvious symmetry of η_l that $\eta_l \subset \zeta_1^\cdot \cap \xi_1^\cdot$. Since, as we have already mentioned, $\zeta_1^\cdot \subset \zeta_1$, we have $\zeta_1^\cdot \cap \xi_1^\cdot \subset \zeta_1 \cap \xi_1$. Now let $(a_1, a_2) \in \zeta_1 \cap \xi_1$. This means that $\xi_1 \langle a_1 \rangle = \xi_1 \langle a_2 \rangle$, and so

$$\xi \langle a_1 a \rangle = \xi \langle a_1 \rangle \cap \xi_2 \langle a \rangle = \xi_1 \langle a_2 \rangle \cap \xi_2 \langle a \rangle = \xi \langle a_2 a \rangle$$

and $a_1 a = a_2 a$ for all $a \in A$, i.e. $(a_1, a_2) \in \eta_l$. The second system of equalities is dual to the first one.

One can easily prove that a pseudosemilattice $(A; \cdot)$ is proper if and only if $(aA)A = aA$ and $A(Aa) = Aa$ for all $a \in A$, i.e. when aA is a principal right and Aa principal left ideals of $(A; \cdot)$ generated by a.

6   BORIS M. SCHEIN

The five identities characterizing proper pseudosemilattices are independent, since the pseudosemilattice $(A; \wedge)$ constructed in Example 1 satisfies all the identities but the last; the pseudosemilattice $(A; \vee)$ from Example 1, all the identities but the fourth; and any idempotent semigroup satisfies the first and the last two identities but, as is known [5], there exist idempotent semigroups satisfying precisely one of the identities $xyxz = xyz$ or $xzyz = xyz$. Finally, any zero semigroup satisfies all the identities but the first.

There are grounds for the conjecture that a pseudosemilattice is associated with a single SDQS if and only if it is an ordinary semilattice.

An algebra with a binary operation satisfying the identity $(xu)(vy) = (xv)(uy)$ we call *entropic* (or *medial*) [6]. Instead of "idempotent semigroup" we will say "band".

PROPOSITION 2. *A pseudosemilattice is associative if and only if it is entropic.*

PROOF. An associative pseudosemilattice satisfies the identities of Lemma 1. It is known [5] that these two identities are equivalent to the property of being entropic. Conversely, every entropic pseudosemilattice, by the entropy identity and the identities of Lemma 1, is associative, since $(xy)z = (xy)(xz) = (xx)(yz) = x(yz)$.

COROLLARY. *The class of associative pseudosemilattices coincides with the class of normal bands.*

The structure of normal bands is completely described in [7] (see also [4]). It would be interesting to describe the structure of arbitrary pseudosemilattices.

Clearly, two structures of normal bands defined on one and the same set of elements coincide if and only if their absorption relatives coincide. This fact generalizes and strengthens an analogous result due to Vagner [8], who considered restrictive semigroups which form a subclass of the class of normal bands.

So normal bands are a natural generalization of semilattices. Note that if the operation of a pseudosemilattice is right or left-*selfdistributive*, i.e. $(xy)z = (xz)(yz)$ or $x(yz) = (xy)(xz)$, then, by Lemma 1, this operation is associative. Normal bands are obviously proper pseudosemilattices, and so all results on such pseudosemilattices are applicable.

The following is an abstract characterization of absorption relatives of normal bands.

THEOREM 3. *A proper pseudosemilattice* $(A; \cdot)$ *with the absorption relative* $(A; \zeta_1, \zeta_2)$ *is a normal band if and only if* $\zeta_2 \circ \zeta_1 = \zeta_1 \circ \zeta_2$ *and, for all* $a_1, a_2 \in A$, *the following identities hold*:

$$\xi_1(\xi_1 \langle a_1 \rangle \cap \xi_2 \langle a_2 \rangle) = \xi_1 \langle a_1 \rangle \cap \xi_1 \circ \xi_2 \langle a_2 \rangle$$

and

$$\xi_2(\xi_1 \langle a_1 \rangle \cap \xi_2 \langle a_2 \rangle) = \xi_2 \circ \xi_1 \langle a_1 \rangle \cap \xi_2 \langle a_2 \rangle,$$

where $\xi_1 = \zeta_1^{-1}$ *and* $\xi_2 = \zeta_2^{-1}$.

PROOF. *Necessity.* Suppose that $(A; \cdot)$ is a normal band and $(a_1, a_2) \in \zeta_2 \circ \zeta_1$, i.e. $(a_1, a) \in \zeta_1$ and $(a, a_2) \in \zeta_2$ for some $a \in A$. Then $aa_1 = a_1$ and $aa_2 = a$; hence $a_1(a_2 a_1) = a_1 a_2 aa_1 = a_1 aa_2 a_1 = a_1 aa_1 = a_1 a_1 = a_1$ and $a_2(a_2 a_1) = a_2 a_1$, i.e. $(a_1, a_2 a_1) \in \zeta_2$ and $(a_2 a_1, a_2) \in \zeta_1$. Therefore $(a_1, a_2) \in \zeta_1 \circ \zeta_2$ and $\zeta_2 \circ \zeta_1 \subset \zeta_1 \circ \zeta_2$. The reverse inclusion is dual to this one.

Now we have

$$\xi_1(\xi_1 \langle a_1 \rangle \cap \xi_2 \langle a_2 \rangle) \subset \xi_1(\xi_1 \langle a_1 \rangle) \cap \xi_1(\xi_2 \langle a_2 \rangle) = \xi_1 \langle a \rangle \cap \xi_1 \circ \xi_2 \langle a_2 \rangle.$$

Suppose that $a \in \xi_1 \langle a_1 \rangle \cap \xi_1 \circ \xi_2 \langle a_2 \rangle$, i.e. $(a, a_1) \in \zeta_1$, $(a, a_0) \in \zeta_1$ and $(a_0, a_2) \in \zeta_2$ for some $a \in A$. Then $a_1 a = a$, $a_0 a = a$ and $a_0 a_2 = a_0$, whence $a(aa_0) = aa_0$, $a_1(aa_0) = aa_0$ and $(aa_0)a_2 = aa_0$, that is $(a, aa_0) \in \xi_1$, $(a_1, aa_0) \in \xi_1$ and $(a_2, aa_0) \in \xi_2$. Therefore $a \in \xi_1 \langle aa_0 \rangle$ and $aa_0 \in \xi_1 \langle a_1 \rangle \cap \xi_2 \langle a_2 \rangle$, i.e. $a \in \xi_1(\xi_1 \langle a_1 \rangle \cap \xi_2 \langle a_2 \rangle)$, which proves the first identity in Theorem 3. The second identity is dual to the first.

Sufficiency. If the absorption relative of a pseudosemilattice $(A; \cdot)$ satisfies the conditions of Theorem 3, then xy is the only element in A which satisfies the condition $\xi \langle xy \rangle = \xi_1 \langle x \rangle \cap \xi_2 \langle y \rangle$. Since ξ_1, ξ_2 and ξ are quasi-order relations and $\xi \subset \xi_1$ and $\xi \subset \xi_2$, we see that $\xi_1 \circ \xi = \xi_1$ and $\xi_2 \circ \xi = \xi_2$. Moreover, $\xi_2 \circ \xi_1 = \xi_1 \circ \xi_2$. Thus

$$\begin{aligned}
\xi \langle (xy)z \rangle &= \xi_1 \langle xy \rangle \cap \xi_2 \langle z \rangle = \xi_1 \circ \xi \langle xy \rangle \cap \xi_2 \langle z \rangle = \xi_1(\xi \langle xy \rangle) \cap \xi_2 \langle z \rangle \\
&= \xi_1(\xi_1 \langle x \rangle \cap \xi_2 \langle y \rangle) \cap \xi_2 \langle z \rangle = \xi_1 \langle x \rangle \cap \xi_1 \circ \xi_2 \langle y \rangle \cap \xi_2 \langle z \rangle \\
&= \xi_1 \langle x \rangle \cap \xi_2 \circ \xi_1 \langle y \rangle \cap \xi_2 \langle z \rangle = \xi_1 \langle x \rangle \cap \xi_2(\xi_1 \langle y \rangle \cap \xi_2 \langle z \rangle) \\
&= \xi_1 \langle x \rangle \cap \xi_2(\xi \langle yz \rangle) = \xi_1 \langle x \rangle \cap \xi_2 \circ \xi \langle yz \rangle \\
&= \xi_1 \langle x \rangle \cap \xi_2 \langle yz \rangle = \xi \langle x(yz) \rangle.
\end{aligned}$$

Since ξ is a quasi-ordered relation, $(xy)z = x(yz)$.

It is easy to verify that in a proper pseudosemilattice $(A; \cdot)$ the fact that ζ_1 and ζ_2 commute means that $(Aa)A = A(aA)$ for all $a \in A$, and the identities formulated in Theorem 3 mean that $Aa_1 \cap A(a_2 A) = A(Aa_1 \cap a_2 A)$ and $(Aa_1)A \cap a_2 A = (Aa_1 \cap a_2 A)A$, respectively.

Let $(A; \cdot)$ be a lower pseudosemilattice associated with an SDQS $(A; \zeta_1, \zeta_2)$. It is easily seen that $(A; \cdot)$ *is a right zero semigroup if and only if* $\zeta_1 = A \times A$ *or, equivalently,* $\zeta_1^* = A \times A$.

The following conditions are equivalent:

(1) $(A; \cdot)$ *is a semilattice.*

(2) *The operation* $\cdot$ *is commutative.*

(3) $\zeta_1^* = \zeta_2^*$.

(4) $\zeta_1 = \zeta_2$.

Indeed, the implications $(4) \to (1) \to (2)$ are obvious. If $\cdot$ is commutative and $(x, y) \in \zeta_1^*$, i.e. $yx = x$, then $xy = x$, i.e. $(x, y) \in \zeta_2^*$ and $\zeta_1^* \subset \zeta_2^*$. Analogously, $\zeta_2^* \subset \zeta_1^*$, so that $\zeta_1^* = \zeta_2^*$. It remains to prove that $(3) \to (4)$. Suppose that $\zeta_1^* = \zeta_2^*$. It follows from the identity $x(xy) = xy$ that $(xy, x) \in \zeta_1^*$, and so $(xy, x) \in \zeta_2^* \subset \zeta_2$. If $y \leq x(\zeta_1)$, then $y \leq y(\zeta_2)$ implies $y \leq xy(\zeta)$. Therefore $y \leq xy(\zeta_2)$. So $y \leq x(\zeta_2)$, i.e. $\zeta_1 \subset \zeta_2$. Analogously $\zeta_2 \subset \zeta_1$.

A semigroup is called a *rectangular band* whenever it satisfies the identity $xyx = x$. Such a semigroup is always both idempotent and normal.

The following conditions are equivalent:

(1) $(A; \cdot)$ is a rectangular band.

(2) $\zeta_1^{\cdot}$ and $\zeta_2^{\cdot}$ are symmetric.

(3) $(\zeta_2 \cap \zeta_2^{-1}) \circ (\zeta_1 \cap \zeta_1^{-1}) = A \times A$.

Indeed, (1) and the identity $xy = x$ imply $yx = yxy = y$, i.e. $\zeta_2^{\cdot} \subset \zeta_2^{\cdot -1}$. Analogously we can prove that $\zeta_1^{\cdot}$ is symmetric. Now let both $\zeta_1^{\cdot}$ and $\zeta_2^{\cdot}$ be symmetric. Then $\eta_l = \zeta_1^{\cdot}$ and $\eta_r = \zeta_2^{\cdot}$. It follows from the identity $x(xy) = xy$ that $(xy, x) \in \zeta_1^{\cdot}$; hence $(x, xy) \in \zeta_1^{\cdot}$. From $(xy)y = xy$ we obtain $(xy, y) \in \zeta_2^{\cdot}$, and so $(x, y) \in \eta_r \circ \eta_l$ and $\eta_r \circ \eta_l = A \times A$. Since $\eta_l = \zeta_1 \cap \zeta_1^{-1}$ and $\eta_r = \zeta_2 \cap \zeta_2^{-1}$, we obtain (3). One can show that the symmetry of $\zeta_i^{\cdot}$ implies transitivity, i.e. this relation is an equivalence.

Let (3) hold. Then $\eta_r \circ \eta_l = A \times A$ and $\eta_l \circ \eta_r = A \times A$, and also

$$\eta_r \cap \eta_l = \zeta_1 \cap \zeta_1^{-1} \cap \zeta_2 \cap \zeta_2^{-1} = \Delta_A,$$

where Δ_A is the identity binary relation on A. So for any two elements $(a_1, a_2) \in A \times A$ there exists a uniquely determined element $a_1 * a_2$ such that $a_1 * a_2 \equiv a_1(\eta_l)$ and $a_1 * a_2 \equiv a_2(\eta_r)$. We claim that $a_1 \wedge a_2 = a_1 * a_2 = a_1 \vee a_2$. Indeed, $a_1 * a_2 \leq a_1(\zeta_1)$ and $a_1 * a_2 \leq a_2(\zeta_2)$, so that $a_1 * a_2 \leq a_1 \wedge a_2(\zeta)$. On the other hand, $a_1 \wedge a_2 \leq a_1(\zeta_1)$ and $a_1 \leq a_1 * a_2(\zeta_1)$; hence $a_1 \wedge a_2 \leq a_1 * a_2(\zeta_1)$. In exactly the same way $a_1 \wedge a_2 \leq a_1 * a_2(\zeta_2)$, and so $a_1 \wedge a_2 \leq a_1 * a_2(\zeta)$ and $a_1 \wedge a_2 = a_1 * a_2$. The equality $a_1 \vee a_2 = a_1 * a_2$ can be checked analogously. It remains to note that $(A; *)$ is both an upper and a lower pseudosemilattice associated with the SDQS $(A; \eta_l, \eta_r)$, and $(A; \eta_l, \eta_r)$ is the absorption relative to $(A; *)$. Therefore $(x * y, x) \in \eta_l$ implies $(x * y) * z = x * z$, and $(x * y, y) \in \eta_r$ implies $x * (y * z) = x * z$. It follows that $*$ is an associative operation and $x * y * x = x * x = x$. Thus $(A; *)$ is a rectangular band.

The following conditions are equivalent for every pseudosemilattice $(A; \cdot)$:

(1) $(A; \cdot)$ satisfies the identity $x(yz) = x(zy)$ [the identity $(xy)z = (yx)z$].

(2) $(A; \cdot)$ satisfies the identity $x(yx) = xy$ [the identity $(xy)x = yx$].

(3) $\zeta_2^{\cdot} = \zeta_1^{\cdot} \circ (\zeta_2 \cap \zeta_2^{-1})(\zeta_1^{\cdot} = \zeta_2^{\cdot} \circ (\zeta_1 \cap \zeta_1^{-1}))$.

(4) $\zeta_2 = \zeta_1 \circ (\zeta_2 \cap \zeta_2^{-1})(\zeta_1 = \zeta_2 \circ (\zeta_1 \cap \zeta_1^{-1}))$.

(5) $(A; \cdot)$ is associative and satisfies condition (2).

Clearly, (5) implies (2). Let (2) hold. Then it follows from $yz \leq y(zy)(\zeta)$ that $yz \leq zy(\zeta_2)$. Interchanging y and z, we obtain $zy \leq yz(\zeta_2)$. Thus $(yz, zy) \in \eta_l$, i.e. $x(yz) = x(zy)$, and so (1) follows from (2). Using the identity in (1), the identities from Lemma 1, and (I) we obtain

$$(xy)x = (xy)((xy)x) = (xy)(x(xy)) = (xy)(xy) = xy,$$
$$(x(yz))z = (x(yz))(xz) = ((xz)(yz))(xz) = (xz)(yz) = x(yz).$$

By Proposition 1, ζ_2^* is transitive, and so $\eta_r \subset \zeta_2^*$ and $\zeta_1^* \circ \eta_r \subset \zeta_2^* \circ \zeta_2^* \subset \zeta_2^*$. Now if $(x, y) \in \zeta_2^*$, i.e. $xy = x$, then $x(yx) = x(xy) = xy = x$ and $(yx)x = yx$; hence $(x, yx) \in \zeta_2^* \cap \xi_2^* = \eta_r$. Since $(yx, y) \in \zeta_1^*$, we have $(x, y) \in \zeta_1^* \circ \eta_r$. Therefore, $\zeta_2^* = \zeta_1^* \circ \eta_r$, i.e. (1) entails (3). If (3) holds, then, by the reflexivity of η_r we have $\zeta_1^* \subset \zeta_1^* \circ \eta_r = \zeta_2^*$, and since $(xy, x) \in \zeta_1^*$ for all $x, y \in A$, we have $(xy, x) \in \zeta_2^*$, i.e. $(xy)x = xy$. If $y \leqslant x(\zeta_1)$, then $y \leqslant y(\zeta_2)$ implies $y \leqslant xy(\zeta)$, which is equivalent to $y \leqslant (xy)x(\zeta)$. Therefore, $y \leqslant (xy)x(\zeta_2)$. However, $(xy)x \leqslant x(\zeta_2)$, and so $y \leqslant x(\zeta_2)$. It follows that $\zeta_1 \subset \zeta_2$.

Now we prove that $\zeta_1 \subset \zeta_2 \to \zeta_2 = \zeta_2^*$ in every pseudosemilattice. Indeed, suppose that $\zeta_1 \subset \zeta_2$ and $x \leqslant y(\zeta_2)$. It follows from $x \leqslant x(\zeta_1)$ that $x \leqslant xy(\zeta)$. On the other hand, $xy \leqslant x(\zeta_1)$. However, $\zeta_1 = \zeta_1 \cap \zeta_2 = \zeta$, and so $xy \leqslant x(\zeta)$ and $xy = x$, i.e. $(x, y) \in \zeta_2^*$. Thus, $\zeta_2 \subset \zeta_2^*$, and it follows that $\zeta_2 = \zeta_2^*$.

In our case $\zeta_2 = \zeta_2^* = \zeta_1^* \circ \eta_r \subset \zeta_2 \circ \eta_r \subset \zeta_2 \circ \zeta_2 = \zeta_2$, i.e. (3) implies (4).

Now let (4) hold. Then $\zeta_1 \subset \zeta_1 \circ \eta_r = \zeta_2$. As we have already proved, $\zeta_2 = \zeta_2^*$. Let $x \leqslant y(\zeta_2)$. Then $(x, z) \in \eta_r$ and $z \leqslant y(\zeta_1)$ for some $y \in A$. Since $z \leqslant z(\zeta_2)$, we obtain $z \leqslant yz(\zeta)$. Since $yz = yx$, we obtain $z \leqslant yx(\zeta)$. Hence $z \leqslant yx(\zeta_2)$. However, $x \leqslant z(\zeta_2)$; hence $x \leqslant yx(\zeta_2)$. This and $x \leqslant x(\zeta_1)$ entail $x \leqslant x(yx)(\zeta)$. Now $x(yx) \leqslant x(\zeta_1)$. Since $\zeta_1 = \zeta_1 \cap \zeta_2 = \zeta$, we see that $x(yx) \leqslant x(\zeta)$. It follows that $x = x(yx)$. We have shown that $x \leqslant y(\zeta_2) \to x = x(yx)$. Now $y(xy) \leqslant y(\zeta_1)$ and $y(xy) \leqslant xy(\zeta_2)$, $xy \leqslant x(\zeta_1)$. It follows from $\zeta_1 \subset \zeta_2$ that $xy \leqslant x(\zeta_2)$. Therefore $y(xy) \leqslant x(\zeta_2)$, and so $y(xy) \leqslant yx(\zeta)$; hence $y(xy) \leqslant yx(\zeta_2)$. Moreover, $xy = (xy)(y(xy)) \leqslant y(xy)(\zeta_2)$, i.e. $xy \leqslant yx(\zeta_2)$. Interchanging x and y, we obtain $yx \leqslant xy(\zeta_2)$, i.e. $(xy, yx) \in \eta_r$. Thus $z(xy) = z(yx)$, i.e. (1) holds. Using this condition and the identities of Lemma 1 together with the identities (I), we get

$$(xy)z = (xy)((xy)z) = (xy)(z(xy)) = (xy)((zy)(xy))$$
$$= (xy)((xy)(zy)) = (xy)(zy) = x(zy) = x(yz),$$

i.e. $(A; \cdot)$ is a band. Thus, (5) holds.

Note that the pseudosemilattice constructed in Example 2 is not associative, even though it satisfies the condition $\zeta_1 \subset \zeta_2$.

Now we turn to pseudolattices. Note at once that *if one of the operations of a pseudolattice is commutative, then the other operation is commutative as well and the pseudolattice is an ordinary lattice*, for the commutativity of one of the operations implies the concidence of the quasi-orders in the SDQS with which the pseudolattice is associated.

If $(A; \wedge, \vee)$ is a pseudolattice, then $(A; \wedge)$ has the absorption relative $(A; \zeta_1^\wedge, \zeta_2^\wedge)$, while $(A; \vee)$ has the converse absorption relative $(A; \xi_1^\vee, \xi_2^\vee)$. If the pseudosemilattice is associated with a SDQS $(A; \zeta_1, \zeta_2)$, then $\zeta_1^\wedge \cup \xi_1^\vee \subset \zeta_1$ and $\zeta_2^\wedge \cup \xi_2^\vee \subset \zeta_2$. It is natural to consider the class of pseudosemilattices such that $\zeta_1^\wedge = \xi_1^\vee$, $\zeta_2^\wedge = \xi_2^\vee$ and both $\zeta_1^\wedge$ and $\zeta_2^\wedge$ are quasi-order relations.

LEMMA 3. *The properties*

$$\zeta_1^\wedge \cap \xi_1^\vee, \qquad \xi_1^\vee \subset \zeta_1^\wedge, \qquad \zeta_2^\wedge \subset \xi_2^\vee, \qquad \xi_2^\vee \subset \zeta_2^\wedge$$

are equivalent to the identities

$$(x \wedge y) \vee x = x, \quad (x \vee y) \wedge x = x, \quad x \vee (y \wedge x) = x, \quad x \wedge (y \vee x) = x$$

respectively.

PROOF. $\zeta_1^\wedge \subset \xi_1^\vee$ means that $x \wedge y = y \to y \vee x = x$. If the latter condition holds, then, substituting $x \wedge y$ for y and using the identities (I) we obtain $x \wedge (x \wedge y) = x \wedge y$, whence $(x \wedge y) \vee x = x$. If the latter identity holds and $x \wedge y = y$, then $y \vee x = (x \wedge y) \vee x = x$. The remaining statements of the lemma follow from the first one by duality principles.

A pseudolattice $(A; \wedge, \vee)$ is called *proper* if the pseudosemilattices $(A; \wedge)$ and $(A; \vee)$ are proper and $\zeta_1^\wedge = \xi_1^\vee$, $\zeta_2^\wedge = \xi_2^\vee$. The absorption relative of the pseudosemilattice $(A; \wedge)$ is called the absorption relative of a proper pseudolattice $(A; \wedge, \vee)$. The pseudolattice constructed in Example 1 is not proper. It satisfies the first and third identities of Lemma 3 but does not satisfy the second and fourth identities, since $(1 \vee 3) \wedge 1 = 2$ and $3 \vee (1 \wedge 3) = 0$.

It follows from Theorem 1 that a proper pseudolattice is associated with its absorption relative. By Theorem 2 and Lemma 3 proper pseudolattices are algebras with two binary operations $\wedge$ and $\vee$ satisfying the identities

$$x \wedge x = x, \qquad x \vee x = x, \qquad (x \wedge y) \wedge (x \wedge z) = (x \wedge y) \wedge z,$$

$$(x \vee y) \vee (x \vee z) = (x \vee y) \vee z, \qquad (x \wedge z) \wedge (y \wedge z) = x \wedge (y \wedge z),$$

$$(x \vee z) \vee (y \vee z) = x \vee (y \vee z), \qquad x \wedge ((x \wedge y) \wedge z) = (x \wedge y) \wedge z,$$

$$x \vee ((x \vee y) \vee z) = (x \vee y) \vee z, \qquad (x \wedge (y \wedge z)) \wedge z = x \wedge (y \wedge z),$$

$$(x \vee (y \vee z)) \wedge z = x \vee (y \vee z)$$

and the four identities from Lemma 3. We have not investigated the interconnections between these fourteen identities.

PROPOSITION 3. *If one of the operations of a proper pseudolattice is associative. so is the other.*

PROOF. By the duality principle we can suppose that the operation $\vee$ of a proper pseudolattice $(A; \wedge, \vee)$ is associative. By Theorem 3 the binary relations $\zeta_1^\vee$ and $\zeta_2^\vee$ commute, so their converses $\zeta_1^\wedge$ and $\zeta_2^\wedge$ commute as well. To prove that $\wedge$ is associative it remains to verify the two identities stated in Theorem 3. We shall prove the first identity only, for the second follows from it by the duality principle. We have seen in the proof of Theorem 3 that the left-hand side of each of these equalities is included into the right-hand side, so it is sufficient to prove that

$$\xi_1^\wedge \langle x \rangle \cap \xi_1^\wedge \circ \xi_2^\wedge \langle y \rangle \subset \xi_1^\wedge (\xi_1^\wedge \langle x \rangle \cap \xi_2^\wedge \langle y \rangle)$$

for all $x, y \in A$.

LEMMA 4. *If $(A; \cdot)$ is a normal band, then $(x, y) \in \zeta_2^{\bullet} \circ \zeta_1^{\bullet} \leftrightarrow x = xyx$.*

PROOF. Clearly, $(x, y) \in \zeta_2^{\bullet} \circ \zeta_1^{\bullet}$ means that $(x, z) \in \zeta_1^{\bullet}$ and $(z, y) \in \zeta_2^{\bullet}$ for some $z \in A$, i.e. $zx = x$ and $zy = z$. Then $x = xx = xzx = xzyx = xyzx = xyx$. If $x = xyx$, then $(x, xy) \in \zeta_1^{\bullet}$ and $(xy, y) \in \zeta_2^{\bullet}$, and so $(x, y) \in \zeta_2^{\bullet} \circ \zeta_1^{\bullet}$.

Returning to our proof of Proposition 3 suppose that $a \in \xi_1^{\wedge} \langle x \rangle \cap \xi_1^{\wedge} \circ \xi_2^{\wedge} \langle y \rangle$. Since $(A; \vee)$ is a normal band, by Lemma 4 we have $(x, a) \in \zeta_1^{\vee}$ and $(y, a)\zeta_1^{\vee} \circ \zeta_2^{\vee}$, or, equivalently, $a \vee x = x$ and $y \vee a \vee y = y$. In the sequel we write ζ_1 and ζ_2 instead of $\zeta_1^{\wedge}$ and $\zeta_2^{\wedge}$ respectively. The pseudolattice $(A; \wedge, \vee)$ is associated with the SDQS $(A; \zeta_1, \zeta_2)$. Clearly $a \leqslant a \vee (x \wedge y)(\zeta_1)$, i.e. $a \in \xi_1 \langle a \vee (x \wedge y) \rangle$. Moreover, $(a \vee (x \wedge y)) \vee x = a \vee x = x$, i.e. $(x, a \vee (x \wedge y)) \in \xi_1$, and also

$$y \vee (a \vee (x \wedge y)) = y \vee y \vee a \vee (x \wedge y) = y \vee a \vee y \vee (x \wedge y)$$
$$= y \vee (x \wedge y) = y,$$

i.e. $(y, a \vee (x \wedge y)) \in \xi_2$. Therefore

$$a \vee (x \wedge y) \in \xi_1 \langle x \rangle \cap \xi_2 \langle y \rangle \quad \text{and} \quad a \in \xi_1(\xi_1 \langle x \rangle \cap \xi_2 \langle y \rangle),$$

i.e.

$$\xi_1 \langle x \rangle \cap \xi_1 \circ \xi_2 \langle y \rangle \subset \xi_1(\xi_1 \langle x \rangle \cap \xi_2 \langle y \rangle).$$

By Theorem 3, $\wedge$ is associative. Proposition 3 is proved.

A proper pseudolattice with associative operations is called *normal*. Thus normal pseudolattices are algebras of the form $(A; \wedge, \vee)$ with two idempotent and associative operations each of which satisfies the normality identity, and the identities of Lemma 3 hold. As for ordinary lattices, one can prove that the first and fourth identities of Lemma 3 imply the idempotence of $\wedge$ and $\vee$.

A pseudolattice $(A; \wedge, \vee)$ is called *rectangular* if the operations $\wedge$ and $\vee$ coincide, i.e. if $x \wedge y = x \vee y$.

PROPOSITION 4. *The following conditions are equivalent for any pseudolattice* $(A; \wedge, \vee)$:
(1) $(A; \wedge, \vee)$ *is rectangular, and both* $(A; \wedge)$ *and* $(A; \vee)$ *are rectangular bands.*
(2) $(A; \wedge)$ *or* $(A; \vee)$ *is a rectangular band.*
(3) *One of the operations* $\wedge$ *or* $\vee$ *satisfies the identities* $(xy)x = x$ *and* $x(yx) = x$.
(4) $(A; \wedge, \vee)$ *is rectangular.*

PROOF. The implications $(1) \to (2) \to (3)$ are obvious. Now suppose that $\wedge$ satisfies (3). Then $(x, x \wedge y) \in \zeta_1^{\wedge}$ and $(x, y \wedge x) \in \zeta_2^{\wedge}$. If $(A; \wedge, \vee)$ is associated with an SDQS $(A; \zeta_1, \zeta_2)$, then $x \leqslant x \wedge y(\zeta_1)$ and $x \leqslant y \wedge x(\zeta_2)$ for all $x, y \in A$. In particular, $y \leqslant x \wedge y(\zeta_2)$, so that $x \vee y \leqslant x \wedge y(\zeta)$. On the other hand, $x \wedge y \leqslant x \vee y(\zeta)$, in every pseudolattice, because $x \wedge y = \leqslant x(\zeta_1)$ and $x \leqslant x \vee y(\zeta_1)$, and also $x \wedge y \leqslant y(\zeta_2)$ and $y \leqslant x \vee y(\zeta_2)$, i.e. $x \wedge y \leqslant x \vee y(\zeta_1)$ and $x \wedge y \leqslant x \vee y(\zeta_2)$. Therefore $x \wedge y = x \vee y$, i.e. (3) implies (4).

Now let (4) hold. Then $x \vee y \leqslant x \wedge y(\zeta)$. Since both $x \leqslant x \vee y(\zeta_1)$ and $y \leqslant x \vee y(\zeta_2)$, we obtain $x \leqslant x \wedge y(\zeta_1)$ and $y \leqslant x \wedge y(\zeta_2)$. It follows from (I) that $(x \wedge y, x) \in \zeta_1$ and $(x \wedge y, y) \in \zeta_2$. Therefore

$$(x, x \wedge y) \in \zeta_1 \cap \zeta_1^{-1} = \eta_l \quad \text{and} \quad (y, x \wedge y) \in \zeta_2 \cap \zeta_2 = \eta_r,$$

i.e. the identities $x \wedge z = (x \wedge y) \wedge z$ and $z \wedge y = z \wedge (x \wedge y)$ hold. Hence $(x \wedge y) \wedge z = x \wedge z = x \wedge (y \wedge z)$, i.e. $\wedge$ is associative. By the corollary to Proposition 2, $(A; \wedge)$ is a normal band. Now the identities of Lemma 3 and condition (1) hold by (4). Proposition 4 is proved.

PROPOSITION 5. *A normal pseudolattice is a lattice if and only if*

$$x \wedge y = x \vee y \rightarrow x = y.$$

PROOF. Let $x \wedge y \wedge x = x$ and $y \wedge x \wedge y = y$. When we were proving the implication $(3) \rightarrow (4)$ in Proposition 4 we saw that if $(x \wedge y) \wedge x = x$ and $y \wedge (x \wedge y) = y$, then $x \wedge y = y \wedge x$. In our case $x \wedge y \wedge x = x$, $y \wedge x \wedge y = y \rightarrow x = y$, i.e. each element of the band $(A; \wedge)$ has a unique generalized inverse [9]. So $\wedge$ is commutative, i.e. $(A; \wedge, \vee)$ is a lattice. The necessity of the condition in Proposition 5 is obvious.

Thus, rectangular pseudolattices and lattices are, in a sense, opposite classes of normal pseudolattices: while the operations $\wedge$ and $\vee$ coincide in the former case, in the latter case they differ as much as it is permitted by the idempotence law.

The *commutativity relation* on an algebra $(A; \cdot)$ is the binary relation $\kappa^{\cdot}$ on A defined as follows:

$$\kappa^{\cdot} = \{(a_1, a_2) \mid a_1 a_2 = a_2 a_1\}.$$

The *inversion relation* of a semigroup is a binary relation

$$\theta^{\cdot} = \{(a_1, a_2) \mid a_1 a_2 a_1 = a_1, \ a_2 a_1 a_2 = a_2\}$$

on A.

Next we prove a series of lemmas which will lead us to Theorem 4 on the structure of normal pseudolattices.

LEMMA 5. *Let $(A; \wedge, \vee)$ be a normal pseudolattice. The inversion relations $\theta^{\wedge}$ and $\theta^{\vee}$ of semigroups $(A; \wedge)$ and $(A; \vee)$ coincide and are a congruence relation on $(A; \wedge, \vee)$, which is denoted by θ and which is called the inversion relation of $(A; \wedge, \vee)$.*

PROOF. It follows from the theory of idempotent semigroups (see [9]) that the inversion relations of bands are congruence relations. It remains to prove that $\theta^{\wedge} = \theta^{\vee}$. By Lemma 4,

$$\theta^{\wedge} = \zeta_2 \circ \zeta_1 \cap \overline{\zeta_2 \circ \zeta_1}^{-1}$$

and, dually,

$$\theta^{\vee} = \zeta_2^{-1} \circ \zeta_1^{-1} \cap \overline{\zeta_2^{-1} \circ \zeta_1^{-1}}^{-1},$$

where $(A; \zeta_1, \zeta_2)$ is the absorption relative of $(A; \wedge, \vee)$. By Theorem 3, ζ_1 and ζ_2 commute, and so $\theta^\wedge = \theta^\vee$.

LEMMA 6. *Let* $(A; \wedge, \vee)$ *be a normal pseudolattice. Then*

$$\zeta \circ \overset{-1}{\zeta} \cap \theta = \overset{-1}{\zeta} \circ \zeta \cap \theta = \Delta_A.$$

PROOF. Clearly, $\Delta_A \subset \zeta \circ \overset{-1}{\zeta} \cap \theta$. Let $(x, y) \in \zeta \circ \overset{-1}{\zeta} \cap \theta$, i.e. $z \leqslant x(\zeta)$ and $z \leqslant y(\zeta)$ for some $z \in A$ and $x \wedge y \wedge x = x$, $y \wedge x \wedge y = y$. Therefore $x = z \vee x = x \vee z$ and $y = z \vee y = y \vee z$, whence

$$x = z \vee x = z \vee x \vee z = z \vee x \vee y \vee x \vee z$$
$$= z \vee x \vee x \vee y \vee z = z \vee x \vee y \vee y \vee z$$
$$= z \vee y \vee x \vee y \vee z = z \vee y \vee z = y \vee z = y.$$

The identity $\overset{-1}{\zeta} \circ \zeta \cap \theta = \Delta_A$ follows dually.

LEMMA 7. *If* $(A; \wedge, \vee)$ *is a normal pseudolattice, then* $\kappa^\wedge = \kappa^\vee = \zeta \circ \overset{-1}{\zeta}$ $\overset{-1}{\zeta} \circ \zeta$, *and* $\kappa = \kappa^\wedge = \kappa^\wedge$ *is a congruence relation on* $(A; \wedge, \vee)$.

We call κ the commutativity relation of $(A; \wedge, \vee)$.

PROOF. Let $(x, y) \in \kappa^\wedge$. Then $x \wedge y \leqslant x(\zeta_1)$ and $x \wedge y = y \wedge x \leqslant x(\zeta_2)$, so that $x \wedge y \leqslant x(\zeta)$. In exactly the same way we have $x \wedge y \leqslant y(\zeta)$. Therefore $(x, x \wedge y) \in \overset{-1}{\zeta}$ and $(x \wedge y, y) \in \zeta$, and so $(x, y) \in \zeta \circ \overset{-1}{\zeta}$. Thus $x^\wedge \subset \zeta \circ \overset{-1}{\zeta}$.

If $(x, y) \in \zeta \circ \overset{-1}{\zeta}$, i.e. if $z \leqslant x(\zeta)$ and $z \leqslant y(\zeta)$ for some $z \in A$, then $z \leqslant x(\zeta_1)$ and $z \leqslant y(\zeta_2)$, so that $z \leqslant x \wedge y(\zeta)$. In the same way, $z \leqslant y \wedge x(\zeta)$. Using the associativity and normality of $\wedge$, we can check straightforwardly that $(x \wedge y, y \wedge x) \in \theta^\wedge$. Therefore

$$(x \wedge y, y \wedge x) \in \zeta \circ \overset{-1}{\zeta} \cap \theta.$$

Hence, by Lemma 6, $x \wedge y = y \wedge x$, i.e. $(x, y) \in \kappa^\wedge$. Thus $\kappa^\wedge \subset \zeta \circ \overset{-1}{\zeta}$. It follows that $\kappa^\wedge = \zeta \circ \overset{-1}{\zeta}$ and, dually, $\kappa^\vee = \overset{-1}{\zeta} \circ \zeta$.

Let $(x, y) \in \kappa^\wedge$. Then

$$x \vee y = ((x \wedge y \vee x) \vee y = (y \wedge x) \vee x \vee y = (y \wedge x) \vee x \vee y \vee y$$
$$= (y \wedge x) \vee y \vee x \vee y = y \vee x \vee y = y \vee x \vee y \vee (x \wedge y)$$
$$= y \vee y \vee x \vee (x \wedge y) = y \vee x \vee (y \wedge x) = y \vee x$$

i.e. $(x, y) \in \kappa^\vee$ and $\kappa^\wedge \subset \kappa^\vee$. Dually, $\kappa^\vee \subset \kappa^\wedge$, and so $\kappa^\wedge = \kappa^\vee$.

Clearly, the binary relation $\kappa = \kappa^\wedge = \kappa^\wedge$ is both reflexive and symmetric. Let $(x, y), (y, z) \in \kappa$. Then

$$x \wedge y \wedge z = (x \wedge y) \wedge (y \wedge z) = (y \wedge x) \wedge (z \wedge y)$$
$$= (y \wedge z) \wedge (x \wedge y) = (z \wedge y) \wedge (y \wedge x)$$
$$= z \wedge y \wedge x.$$

Therefore

$$x \wedge (x \wedge y \wedge z) = x \wedge y \wedge z$$

and

$$(x \wedge y \wedge z) \wedge x = (z \wedge y \wedge x) \wedge x = z \wedge y \wedge x = x \wedge y \wedge z,$$

i.e. $x \wedge y \wedge z \leqslant z(\zeta)$. Analogously $x \wedge y \wedge z \leqslant z(\zeta)$. So

$$(x, z) \in \zeta \circ \overset{-1}{\zeta} = \kappa,$$

i.e. κ is transitive. Therefore κ is an equivalence relation. If $(x, z) \in \kappa$, then

$$(x \wedge z) \wedge (y \wedge z) = (x \wedge y) \wedge (z \wedge z) = (y \wedge x) \wedge (z \wedge z)$$
$$= (y \wedge z) \wedge (x \wedge z),$$

i.e. $(x \wedge z, y \wedge z) \in \kappa$. In exactly the same way, $(z \wedge x, z \wedge y) \in \kappa$ for every $z \in A$. It follows that κ is stable, i.e. κ is a congruence relation on $(A; \wedge)$. Dually κ is a congruence on $(A; \vee)$.

THEOREM 4. *An algebra is a normal pseudolattice if and only if it is decomposable into a direct product of a lattice and a rectangular pseudolattice. For every normal pseudolattice $\mathfrak{A}$ the only (up to isomorphism) decomposition into a direct product of a lattice and a rectangular pseudolattice is $(\mathfrak{A}/\theta) \times (\mathfrak{A}/\kappa)$.*

PROOF. Since lattices and rectangular pseudolattices are normal pseudolattices, their direct products are also normal pseudolattices, because the class of all normal pseudolattices is defined by identities.

Let $\mathfrak{A} = (A; \wedge, \vee)$ be a normal pseudolattice. By Lemmas 5–7, θ and κ are congruence relations, and $\theta \cap \kappa = \Delta_A$. Let $x, y \in A$ and let $a = (y \wedge x \wedge y) \vee x \vee (y \wedge x \wedge y)$. Then it is easy to verify that $a \vee x \vee a = a$. Moreover,

$$x \vee a \vee x = x \vee (y \wedge x \wedge y) \vee x = (x \wedge y) \vee x \vee (y \wedge x \wedge y) \vee x$$

$$= (x \wedge y) \vee (y \wedge (x \wedge y)) \vee x \vee x$$

$$= (x \wedge y) \vee x \vee x = x,$$

so that $(x, a) \in \theta$. Clearly, $(y \wedge x \wedge y) \vee a = a \vee (y \wedge x \wedge y) = a$, i.e. $(y \wedge x \wedge y, a) \in \overset{-1}{\zeta}$. It is also clear that $(y \wedge x \wedge y, y) \in \zeta$, and so $(a, y) \in \zeta \circ \overset{-1}{\zeta} = \kappa$. Therefore $(x, y) \in \kappa \circ \theta$. Consequently, $\kappa \circ \theta = A \times A$. Together with $\theta \cap \kappa = \Delta_A$ this means [10] that $\mathfrak{A}$ is decomposed in a direct product of quotient algebras $\mathfrak{A}/\theta$ and $\mathfrak{A}/\kappa$, which are normal pseudolattices, since a homomorphic image of a normal pseudolattice is a normal pseudolattice.

It is known (see [9]) that if $(A; \cdot)$ is a band, then the quotient semigroup $(A; \cdot)/\theta^{\cdot}$ is commutative. So the operations in the quotient algebra $\mathfrak{A}/\theta$ are commutative, i.e. $\mathfrak{A}/\theta$ is a lattice. Clearly, $\zeta \subset \zeta \circ \overset{-1}{\zeta} \subset \kappa$. Therefore $(x \wedge y, x \vee y) \in \zeta \subset \kappa$, and so

$$\kappa \langle x \rangle \wedge \kappa \langle y \rangle = \kappa \langle x \wedge y \rangle = \kappa \langle x \vee y \rangle = \kappa \langle x \rangle \vee \kappa \langle y \rangle,$$

where $\kappa \langle x \rangle$ is the κ-class containing an element $x \in A$, and also $\kappa \langle x \rangle \wedge \kappa \langle y \rangle$ is the result of application of the operation of $\mathfrak{A}/\kappa$ to the elements $\kappa \langle x \rangle$ and $\kappa \langle y \rangle$ of this algebra. Thus, the both operations of $\mathfrak{A}/\kappa$ coincide, i.e. $\mathfrak{A}/\kappa$ is a rectangular pseudolattice.

To prove the uniqueness of decomposition suppose that $\mathfrak{A}$ is isomorphic to $(\mathfrak{A}/\alpha) \times (\mathfrak{A}/\beta)$, where $\mathfrak{A}/\alpha$ is a lattice and $\mathfrak{A}/\beta$ a rectangular pseudolattice. It is known (see [9]) that θ is the smallest congruence on an idempotent semigroup such that the quotient semigroup is commutative. Therefore $\theta \subset \alpha$. Now let $(x, y) \in \kappa$. Then

$$\beta \langle x \rangle = \beta \langle x \rangle \wedge \beta \langle y \rangle \wedge \beta \langle x \rangle = \beta \langle x \wedge y \wedge x \rangle$$
$$= \beta \langle y \wedge x \wedge y \rangle = \beta \langle y \rangle \wedge \beta \langle x \rangle \wedge \beta \langle y \rangle = \beta \langle y \rangle,$$

i.e. $(x, y) \in \beta$. Thus $\kappa \subset \beta$. It follows that the natural homomorphisms of $\mathfrak{A}/\theta$ onto $\mathfrak{A}/\alpha$ and of $\mathfrak{A}/\kappa$ onto $\mathfrak{A}/\beta$ are isomorphisms. Theorem 4 is proved.

COROLLARY. *Let A and B be nonempty sets, and let $(L; \wedge, \vee)$ be a lattice. Introduce the following binary operations $\wedge$ and $\vee$ on the Cartesian product $A \times L \times B$ of the sets A, L, and B:*

$$(a_1, l_1, b_1) \wedge (a_2, l_2, b_2) = (a_1, l_1 \wedge l_2, b_2),$$
$$(a_1, l_1, b_1) \vee (a_2, l_2, b_2) = (a_1, l_1 \vee l_2, b_2).$$

Then $A \times L \times B$ is a normal pseudolattice, and every normal pseudolattice can be obtained in this way.

Let a set A be endowed with a pair (ε, η) of commuting equivalence relations. A subset $\mathfrak{a} \subset A$ is called *rectangular* if $\varepsilon(\mathfrak{a}) \cap \eta(\mathfrak{a}) = \mathfrak{a}$, where $\varepsilon(\mathfrak{a})$ is the union of all those ε-classes that have a nonempty intersection with $\mathfrak{a}$. It is easy to verify that if $\mathfrak{a}_1$ and $\mathfrak{a}_2$ are arbitrary subsets of A, then $\varepsilon(\mathfrak{a}_1) \cap \eta(\mathfrak{a}_2)$ is rectangular. Introduce an operation $\cdot$ on the set $\mathfrak{R}(A)$ of all rectangular subsets as follows: $\mathfrak{a}_1 \cdot \mathfrak{a}_2 = \varepsilon(\mathfrak{a}_1) \cap \eta(\mathfrak{a}_2)$. Then $(\mathfrak{R}(A); \cdot)$ is a normal band. Subbands of this band are called *special*. Clearly, every special band is normal.

THEOREM 5. *Every normal band is isomorphic to a special band.*

PROOF. Let $(A; \cdot)$ be a normal band. With every element $a \in A$ we associate a set $\overset{-1}{\zeta} \langle a \rangle$. Since ζ is an order relation, this correspondence is one-to-one. We omit the easy proof that the subsets $\overset{-1}{\zeta} \langle a \rangle$ are rectangular relative to $(A; \eta_l, \eta_r)$ (note that η_l and η_r are Green equivalences of the semigroup $(A; \cdot)$, and so they commute). The correspondence $a \to \zeta^{-1} \langle a \rangle$ is an isomorphism of $(A; \cdot)$ onto a special band.

An analogous representation theorem can be proved for distributive pseudolattices. Other representations for normal bands are constructed in [11].

BIBLIOGRAPHY

1. P. Jordan, *Halbgruppen von Idempotenten und nichtkommutative Verbände*, J. Reine Angew. Math. **211** (1962), 136–161.

2. ______, *The mathematical theory of quasi order, semi groups of idempotents and noncommutative lattices—a new field of modern algebra*, Report, No. ARL228, Part I, U.S. Air Force Office of Aerospace Research, European Office (Univ. of Hamburg), 1961.

3. Max Dieter Gerhardts, *Schrägverbände und Quasiordnungen*, Math. Ann. **181** (1969), 65–73.

4. B. M. Šaĭn [Schein], *Entropic bands and noncommutative semilattices*, Eighteenth Hercen Readings Math. (1965), Program and Abstracts of Reports, Leningrad. Gos. Ped. Inst., Leningrad, 1965, pp. 11–13. (Russian)

5. Naoki Kimura, *Note on idempotent semigroups*. III, Proc. Japan Acad. **34** (1958), 113–114.

6. Albert Sade, *Quasigroups obéissant à certaines lois*, Rev. Fac. Sci. Univ. Istanbul Sér. A **22** (1957), 151–184.

7. Naoki Kimura, *The structure of idempotent semigroups*. I, Pacific J. Math. **8** (1958), 257–275.

8. V. V. Vagner, *Restrictive semigroups*, Izv. Vysš. Učebn. Zaved. Matematika **1962**, no. 6(31), 19–27. (Russian)

9. B. M. Šaĭn [Schein], *On the theory of inverse semigroups and generalized grouds*, Theor. Semigroups and Appl., No. 1, Izdat. Saratov. Univ., Saratov, 1965, pp. 286–324; English transl. in Amer. Math. Soc. Transl. (2) **113** (1979).

10. P. M. Cohn, *Universal algebra*, Harper & Row, New York, 1965.

11. Mario Petrich, *Homomorphisms of a semigroup onto normal bands*, Acta Sci. Math. (Szeged) **27** (1966), 185–196.

Translated by B. M. SCHEIN

Amer. Math. Soc. Transl.
(2) Vol. **119**, 1983

On the Representation of a Large Even Integer
As a Sum of a Prime
and the Product of At Most Three Primes*

HSIEH SHENG-KANG

§1. The result of this paper

M. B. Barban [1], Wang Yuan [2], and Pan Cheng-Tung [3] proved that any sufficiently large even integer can be expressed as the sum of a prime and the product of at most four primes. Under the generalized Riemann hypothesis, Wang Yuan [2] proved that any sufficiently large even integer can be represented as the sum of a prime and the product of at most three primes.

The purpose of this paper is to give a new proof of the above result:

THEOREM. *Any sufficiently large even integer can be represented as the sum of a prime and the product of at most three primes.*

§2. The lower bound of $P_1(x, x^{1/7})$

Assume that x is a sufficiently large even integer and that $2 < \xi < x$ is real. Let $P = \prod_{2 < p \leqslant \xi} p$, $a_n = \log n e^{-n \log x / x}$, and $P_1(x, \xi) = \Sigma_{(x-p,P)=1; 2 < p \leqslant x} a_p$.

LEMMA 1. $P_1(x, x^{1/7}) > 8.149967 c_{2x} x / \log^2 x$, *where c_{2x} is the same as in the definition of* [4].

PROOF. Similarly to [3], we know that

$$P_1(x, x^{1/18}) > 36(1 - 0.00732)c_{\gamma,x} x / \log^2 x = 35.73648 \, c_{\gamma,x} x / \log^2 x,$$

where

$$c_{\gamma,x} = e^{-\gamma} \prod_{p>2} \left(1 - \frac{1}{(p-1)^2}\right) \prod_{\substack{p|x \\ p>2}} \frac{p-1}{p-2}.$$

1980 *Mathematics Subject Classification.* Primary 10J15.
*Translation of Shuzue Jinzhan **8** (1965), 209–216; MR **37** #5171.

From [3] we see that

$$P_1(x, x^{1/7}) > \left(35.73648 - \int_8^{18} \frac{2e^\gamma \, du}{(3u-8)/8 - 1 - ((3u-8)/16)\log((3u-8)/16)} \right.$$

$$\left. - \int_7^8 \frac{32 e^\gamma \, du}{3u - 8} \right) \frac{c_{\gamma, x} x}{\log^2 x}.$$

Through computation we now obtain the lemma.

§3. The function $\rho(x, k, l)$ and the Selberg sieve method

Assuming $k > l \geq 0$, let $\rho(x, k, l) = \sum_{n \equiv l(\mathrm{mod}\ k); n \leq x} a_n$ and $\rho(x) = \rho(x, l, 0) = \sum_{n \leq x} a_n$.

It is easy to prove that $\rho(x) = x + O((x \log \log x)/\log x)$.

LEMMA 2. $\rho(x, k, l) = \frac{1}{k}\rho(x) + O(\log x)$ *uniformly for* $k > l \geq 0$; *here the constants relating to "O" do not depend on x, k or l.*

PROOF. Assume that $t_0 \log t_0 - x/\log x = 0$. It is easy to see that a_n is monotonically increasing in the interval (l, t_0) and monotonically decreasing in (t_0, x). Thus we have

$$-\log x < \sum_{\substack{n \equiv l(\mathrm{mod}\ k) \\ n \leq t_0}} a_n - \sum_{\substack{n \equiv 0(\mathrm{mod}\ k) \\ n \leq t_0}} a_n < \log x$$

and

$$-\log x < \sum_{\substack{n \equiv l(\mathrm{mod}\ k) \\ r_0 < n \leq x}} a_n - \sum_{\substack{n \equiv 0(\mathrm{mod}\ k) \\ t_0 < n \leq x}} a_n < \log x.$$

Putting the two formulas together, we obtain

$$\rho(x, k, 0) = \rho(x, k, l,) + O(\log x).$$

Hence

$$k\rho(x, k, 0) = \sum_{l=0}^{k-1} \rho(x, k, l) + O(k \log x) = \rho(x) + O(k \log x).$$

Therefore

$$\rho(x, k, l) = \rho(x, k, 0) + O(\log x) = \frac{1}{k}\rho(x) + O(\log x).$$

Let x be a sufficiently large even integer, and let ξ_1 and ξ_2 $(2 < \xi_1 < \xi_2 < x)$ be two real numbers. Fix a set of positive integers

$$(\omega) \qquad\qquad\qquad y; q, a; a', b'; K$$

such that $2 \leq y \leq x$, $q < \xi_1$, $2 \mid q$, $q = O(1)$, and when $p \leq \xi_2$, if $p \mid y$, then $a' \equiv b'$ (mod p), while if $p \nmid y$, then $a' \not\equiv b'$ (mod p); and let $\mu(K) \neq 0$ for $K < x$. Let $\pi \xi = \prod_{p \nmid qK; p \leq \xi} p$ and

$$P_\omega(x, \xi_1, \xi_2) = \sum_{\substack{(n-a', \pi_{\xi_1})=1 \\ (n-b', \pi_{\xi_2})=1 \\ n \leq x, n \equiv a(\mathrm{mod}\ q)}} a_n.$$

LEMMA 3. *If $c > 0$, then*

$$P_\omega(x, \xi_1, \xi_2) < \rho(x)\Big/ q \sum_{\substack{n|\pi_{\xi_2} \\ n \leqslant \xi_1^c}} \frac{\mu(n)}{f(n)} + O(\xi^{2c} \log^7 x)$$

uniformly with respect to (ω). Here, when $n \mid \pi_{\xi_2}$,

$$f(n) = \prod_{p|n} f(p), \qquad f(p) = \frac{1}{g(p)} - 1; \qquad g(1) = 1,$$

$$g(p) = 2/p \quad (p \nmid y, \, p < \xi_1), \qquad g(p) = 1/p \quad (p \mid y \text{ or } \xi_1 < p).$$

PROOF. If $k \mid \pi_{\xi_2}$, then we can write $k = k_1 k_2$, with $k_1 \mid \pi_{\xi_1}$ and $(k_2, \pi_{\xi_1}) = 1$. It is easy to see that the system of congruences

$$\begin{cases} (n - a')(n - b') \equiv 0 & (\mathrm{mod} \ k_1) \\ n - b' \equiv 0 & (\mathrm{mod} \ k_2) \\ n \equiv a & (\mathrm{mod} \ q) \end{cases}$$

has $2^{\Omega(k_1) - \Omega((k_1, y))}$ solutions in the interval $1 \leqslant n \leqslant kq$; therefore from Lemma 1 we see that

$$\sum_{\substack{k_1|(n-a')(n-b') \\ k_2|n-b' \\ n \leqslant x, \, n \equiv a(\mathrm{mod} \ q)}} a_n = \frac{\rho(x)}{kq} q^{\Omega(k_1) - \Omega((k_1, y))} + O\big(2^{\Omega(k_1)} \log x\big).$$

When $k \mid \pi_{\xi_2}$, let

$$\lambda_k = \frac{\mu(k)}{g(k)f(k)} \sum_{\substack{n \leqslant \xi_1^c/k \\ (n,k)=1 \\ n|\pi_{\xi_2}}} \frac{\mu^2(n)}{n} \Big/ \sum_{\substack{l \leqslant \xi_1^c \\ l|\pi_{\xi_2}}} \frac{\mu^2(l)}{f(l)}.$$

Since $\lambda_1 = 1$ and $\lambda_d = 0$ for $d > \xi_1^c$, we have

$$P_\omega(x, \xi_1, \xi_2) = \sum_{\substack{(n-a', \pi_{\xi_1})=1 \\ (n-b', \pi_{\xi_2})=1 \\ n \leqslant x, \, n \equiv a(\mathrm{mod} \ q)}} a_n = \sum_{\substack{(n-a', \pi_{\xi_1})=1 \\ (n-b', \pi_{\xi_2})=1 \\ n \leqslant x, \, n \equiv a(\mathrm{mod} \ q)}} a_n \left(\sum_{\substack{d_1|((n-a')(n-b'), \pi_{\xi_1}) \\ d_2|(n-b', \pi_{\xi_2}/\pi_{\xi_1})}} \lambda_{d_1 d_2} \right)^2$$

$$\leqslant \sum_{\substack{n \leqslant x, \, n \equiv a(\mathrm{mod} \ q)}} a_n \left(\sum_{\substack{d_1|((n-a')(n-b'), \pi_{\xi_1}) \\ d_2|(n-b', \pi_{\xi_2}/\pi_{\xi_1})}} \lambda_{d_1 d_2} \right)^2$$

$$= \sum_{\substack{d_1 d_2 \leqslant \xi_1^c \\ d_1|\pi_{\xi_1} \\ d_2|\pi_{\xi_2}/\pi_{\xi_1}}} \sum_{\substack{d_1' d_2' \leqslant \xi_1^c \\ d_1'|\pi_{\xi_1} \\ d_2'|\pi_{\xi_2}/\pi_{\xi_1}}} \lambda_{d_1 d_2} \lambda_{d_1' d_2'} \sum_{\substack{n \leqslant x, \, n \equiv a(\mathrm{mod} \ q) \\ \{d_1, d_1'\}|(n-a')(n-b') \\ \{d_2, d_2'\}|n-b'}} a_n$$

$$= \frac{\rho(x)}{q} \sum_{\substack{d_1 d_2 \leqslant \xi_1^c \\ d_1 | \pi_{\xi_1} \\ d_2 | \pi_{\xi_2}/\pi_{\xi_1}}} \sum \; \sum_{\substack{d_1' d_2' \leqslant \xi_1^c \\ d_1' | \pi_{\xi_1} \\ d_2' | \pi_{\xi_2}/\pi_{\xi_1}}} \sum \lambda_{d_1 d_2} \lambda_{d_1' d_2'} g(\{d_1 d_2, d_1' d_2'\})$$

$$+ O\left(\sum_{\substack{d \leqslant \xi_1^c \\ d | \pi_{\xi_2}}} \sum_{\substack{d' \leqslant \xi_1^c \\ d' | \pi_{\xi_2}}} |\lambda_d \lambda_{d'}| \, 2^{\Omega(d) + \Omega(d')} \log x \right)$$

$$= \frac{\rho(x)}{q} \sum_{\substack{d \leqslant \xi_1^c \\ d | \pi_{\xi_2}}} \sum_{\substack{d' \leqslant \xi_1^c \\ d' | \pi_{\xi_2}}} \lambda_d \lambda_{d'} g(\{d, d'\}) + R = \frac{\rho(x)}{q} Q + R,$$

where $\{d, d'\}$ denotes the least common multiple of d and d'. The rest of the proof is the same as in [4].

§4. The application of Lemma 3

Let $\xi_1 = x^{1/2c} \log^{-6} x$ and $\xi_2 = x^{1/2}$. Write $d = 2c$.

1°. If $1 \leqslant c \leqslant 3$, then similarly to [4] we have

$$\sum_{\substack{n \leqslant \xi_1^c \\ n | \pi_{\xi_2}}} \frac{\mu^2(n)}{f(n)} \geqslant \sum_{\substack{n \leqslant \xi_1^c \\ n | \pi_{\xi_1}}} \frac{\mu^2(n)}{f(n)} + \sum_{\xi_1 < p \leqslant \xi_1^c} \frac{1}{p-1} \sum_{\substack{n \leqslant \xi_1^c/p \\ n | \pi_{\xi_1}}} \frac{\mu^2(n)}{f(n)}$$

$$\geqslant \sum_{\substack{n \leqslant \xi_1^c \\ (n, qK) = 1}} \frac{\mu^2(n)}{f(n)} - \sum_{\substack{\xi_1 < p \leqslant \xi_1^c \\ p \nmid K}} \frac{1}{p} \sum_{\substack{n \leqslant \xi_1^c/p \\ (n, qK) = 1}} \frac{\mu^2(n)}{f(n)} + O\left(\frac{\log^6 x}{x^{1/d}} \right)$$

$$= \frac{c'}{2} \left[(d-1)^2 - \frac{d^2}{2} \log \frac{d}{2} + \frac{d^2}{4} \right] \log^2 \xi_1 + O\left(\prod_{\substack{p | qKy \\ p > 2}} \frac{p-2}{p-1} \cdot \frac{\log^2 x}{\log \log x} \right),$$

where

$$c' = \frac{1}{4} \prod_{p | qK} \frac{p-1}{p} \prod_{p > 2} \frac{(p-1)^2}{p(p-2)} \prod_{\substack{p | qKy \\ p > 2}} \frac{p-2}{p-1}.$$

2°. From 1° we immediately obtain

LEMMA 4. *If $2 \leqslant d \leqslant 6$, then*

$$P_\omega(x, x^{1/d}, x^{1/2}) < c_{qKy} \Lambda(d)(x/\log^2 x) + O(c_{qKy} x/\log^2 x \log \log x),$$

uniformly with respect to (ω), where

$$c_{qKy} = \frac{2e^{-2\gamma}}{q} \prod_{p > 2} \left(1 - \frac{1}{(p-2)^2} \right) \prod_{p | qK} \frac{p}{p-1} \prod_{\substack{p | qKy \\ p > 2}} \frac{p-1}{p-2},$$

$$\Lambda(d) = \frac{4e^{2\gamma} d^2}{(d-1)^2 + d^2/4 - (d^2/2) \log(d/2)}.$$

It is easy to show that $\Lambda(d)/d^2$ is a decreasing function in the interval $2 \leqslant d \leqslant 6$.

§5. The set $\mathfrak{M}$

Let $\mathfrak{M}$ be a set of integers satisfying the following conditions:

$$x^{1/2} < n \leqslant x, \quad 2 \mid n, \quad n(x - n) \not\equiv 0 \pmod{p} \quad (2 < p \leqslant x^{1/7}),$$

$$n \not\equiv 0 \pmod{p}(x^{1/7} < p \leqslant x^{1/2}), \quad x - n \not\equiv 0 \pmod{p^2} \quad (x^{1/7} < p \leqslant x^{1/2}).$$

Obviously $\sum_{n \in \mathfrak{M}} a_n = P_1(x, x^{1/7}) + O(x^{1/2} \log x)$.

Let $J(1, 3)$ denote the set of all primes p such that $x - p$ $(1 < p < x)$ has at most three prime factors.

LEMMA 5.

$$P_1(x, x^{1/7}) < \sum_{p \in J(1,3)} a_p + \frac{1}{6} \sum_{\substack{p'p'' \leqslant x^{1/2} \\ p',p'' > x^{1/7} \\ (p'p'',x)=1}} P_{\omega_{p'p''}}\left(\frac{x}{p'p''}, x^{1/7}, \left(\frac{x}{p'p''} \right)^{1/2} \right)$$

$$+ O(x^{6/7+\varepsilon}).$$

Here p' and p'' denote primes, and $(\omega_{p'p''})$ depends on p' and p''.

PROOF. Let $p'' > p' > x^{1/7}$ $(p'p'' \leqslant x^{1/2})$ be two primes, and let $\Gamma_{p'p''}$ denote the set of elements n in $\mathfrak{M}$ satisfying the congruence $x - n \equiv 0 \pmod{p'p''}$. Let $a_{p'p''}$ $(0 < a_{p'p''} \leqslant p'p'')$ be the solution of the above congruence. Then $n = mp'p'' + a_{p'p''}$.

When $(p'p'', x) = 1$, the system of congruences

$$mp'p'' + a_{p'p''} \equiv x \pmod{p} \quad (2 < p \leqslant x^{1/7})$$

has a unique solution $\tilde{a}'$ $(0 < \tilde{a}' \leqslant \pi_{x^{1/7}})$. If we take $K = p'p''$, the system of congruences

$$mp'p'' + a_{p'p''} \equiv 0 \pmod{p} \quad (p \mid \pi_{x^{1/2}})$$

has a unique solution $\tilde{b}'$ $(0 < \tilde{b}' \leqslant \pi_{x^{1/2}})$.

Let $\tilde{a}$ be the solution of $mp''p' + a_{p'p''} \equiv 1 \pmod{2}$. Define a set of positive integers

$$(\omega_{p'p''}) \qquad\qquad y = x; q = 2, \tilde{a}; \tilde{a}', \tilde{b}'; K = p'p''.$$

Now we have

$$\sum_{n \in \Gamma_{p'p''}} a_n \leqslant P_{\omega_{p'p''}}\left(\frac{x}{p'p''}, x^{1/7}, x^{1/2} \right) \leqslant P_{\omega_{p'p''}}\left(\frac{x}{p'p''}, x^{1/7}, \left(\frac{x}{p'p''} \right)^{1/2} \right).$$

When $(p'p'', x) > 1$, obviously

$$\sum_{n \in \Gamma_{p'p''}} a_n < x^{6/7} \log x.$$

Since $p'' > p' > x^{1/7}$, we have

$$\sum_{(p'p'',x)>1} \sum_{n\in\Gamma_{p'p''}} a_n = O\left(x^{6/7}\log x\right).$$

Therefore

$$\sum_{\substack{p'p''\leqslant x^{1/2}\\ p',p''>x^{1/7}}}\sum \sum_{n\in\Gamma_{p'p''}} a_n \leqslant \sum_{\substack{p'p''\leqslant x^{1/2}\\ p',p''>x^{1/7}\\ (p'p'',x)=1}}\sum P_{\omega_{p'p''}}\left(\frac{x}{p'p''}, x^{1/7}, \left(\frac{x}{p'p''}\right)^{1/2}\right) + O\left(x^{6/7}\log x\right).$$

If $p \in \mathfrak{M}$ and $x - p$ has at least four prime factors between $x^{1/7}$ and $x^{1/2}$, let them be $q_1 < q_2 < q_3 < q_4 < \ldots$; there are six pairs (q_μ, q_ν) $(1 \leqslant \mu < \nu \leqslant 4)$. If the product of one of the pairs is greater than $x^{1/2}$, then the product of the other two primes must be less than $x^{1/2}$. Hence at least three pairs (q_μ, q_ν) $(1 \leqslant \mu < \nu \leqslant 4)$ satisfy $q_\mu q_\nu \leqslant x^{1/2}$. Thus p must belong to at least three different $\Gamma_{p'p''}$.

If $p \in \mathfrak{M}$ and $x - p$ has exactly three prime factors between $x^{1/7}$ and $x^{1/2}$, then when the product of these three prime factors is greater than $x^{1/2}$ we have $p \in J(1, 3)$, and when the product of these three prime factors is less than $x^{1/2}$, p belongs to three different $\Gamma_{p'p''}$.

If $p \in \mathfrak{M}$ and $x - p$ has at most two prime factors between $x^{1/7}$ and $x^{1/2}$, then $p \in J(1, 3)$.

Hence, if $p \in \mathfrak{M}$, then p must belong to three different $\Gamma_{p'p''}$ or belong to $J(1, 3)$. From this we have

$$P_1(x, x^{1/7}) \leqslant \sum_{p\in J(1,3)} a_p + \frac{1}{3} \sum_{\substack{p'p''\leqslant x^{1/2}\\ p''>p'>x^{1/7}}}\sum \sum_{n\in\Gamma_{p'p''}} a_n + O\left(x^{1/2}\log x\right)$$

$$\leqslant \sum_{p\in J(1,3)} a_p + \frac{1}{6} \sum_{\substack{p'p''\leqslant x^{1/2}\\ p',p''>x^{1/7}\\ (p'p'',x)=1}}\sum\sum P_{\omega_{p'p''}}\left(\frac{x}{p'p''}, x^{1/7}, \left(\frac{x}{p'p''}\right)^{1/2}\right) + O\left(x^{6/7+\varepsilon}\right).$$

§6. Proof of the theorem

LEMMA 6. *Let there be given two real numbers, $\beta > 0$ and $\alpha > 2$. Then*

$$\sum_{\substack{p',p''>x^{1/\beta}\\ p'p''\leqslant x^{\alpha/\beta}}}\sum \frac{1}{p'p''} = \log\alpha\log(\alpha-1)$$

$$- \sum_{n=1}^{\infty} \frac{1}{n^2}\left(\left(\frac{\alpha-1}{\alpha}\right)^n - \left(\frac{1}{\alpha}\right)^n\right) + O(\log^{-1/2} x).$$

PROOF. Take $\delta = \log^{-1/2} x$ and $M = [(\alpha - 2)/\delta\beta]$. Then

$$\sum_{\substack{p',p''>x^{1/\beta} \\ p'p''\leqslant x^{\alpha/\beta}}} \frac{1}{p'p''} = \sum_{x^{1/\beta}<p'<x^{(\alpha-1)/\beta}} \frac{1}{p'} \sum_{x^{1/\beta}<p''\leqslant x^{\alpha/\beta}/p'} \frac{1}{p''}$$

$$\leqslant \sum_{n=0}^{M} \sum_{x^{1/\beta+n\delta}<p'<x^{1/\beta+(n+1)\delta}} \frac{1}{p'} \sum_{x^{1/\beta}<p''\leqslant x^{(\alpha-1)/\beta-n\delta}} \frac{1}{p''}$$

$$= \sum_{n=0}^{M} \log(\alpha - 1 - n\beta\delta) \int_{1+n\beta\delta}^{1+(n+1)\beta\delta} \frac{dt}{t} + O(\log^{-1/2} x)$$

$$= \sum_{n=0}^{M} \int_{1+n\beta\delta}^{1+(n+1)\beta\delta} \frac{\log(\alpha - t)}{t} dt + O(\log^{-1/2} x) + O\left(\sum_{n=0}^{M} \delta \int_{1+n\beta\delta}^{1+(n+1)\beta\delta} \frac{dt}{t} \right)$$

$$= \int_{1}^{\alpha-1} \frac{\log(\alpha - t)}{t} dt + O(\log^{-1/2} x)$$

$$= \log\alpha \int_{1}^{\alpha-1} \frac{dt}{t} + \int_{1}^{\alpha-1} \frac{\log(1 - t/\alpha)}{t} dt + O(\log^{-1/2} x)$$

$$= \log\alpha \log(\alpha - 1) - \int_{1}^{\alpha-1} \frac{1}{t} \sum_{n=1}^{\infty} \frac{t^n}{n\alpha^n} dt + O(\log^{-1/2} x)$$

$$= \log\alpha \log(\alpha - 1) - \sum_{n=1}^{\infty} \frac{1}{n^2} \left(\left(\frac{\alpha - 1}{\alpha}\right)^n - \left(\frac{1}{\alpha}\right)^n \right) + O(\log^{-1/2} x).$$

Similarly we can prove that

$$\sum_{\substack{p',p''>x^{1/\beta} \\ p'p''\leqslant x^{\alpha/\beta}}} \frac{1}{p'p''} \geqslant \log\alpha \log(\alpha - 1)$$

$$- \sum_{n=1}^{\infty} \frac{1}{n^2} \left(\left(\frac{\alpha - 1}{\alpha}\right)^n - \left(\frac{1}{\alpha}\right)^n \right) + O(\log^{-1/2} x).$$

Thus we obtain the lemma.

LEMMA 7. *When x is sufficiently large,*

$$\sum_{\substack{p',p''>x^{1/7} \\ p'p''\leqslant x^{1/2} \\ (p'p'',x)=1}} P_{\omega_{p'p''}}\left(\frac{x}{p'p''}, x^{1/7}, \left(\frac{x}{p'p''}\right)^{1/2} \right) < 45.7977 \frac{c_{2x}x}{\log^2 x},$$

where c_{2x} is as defined in [4].

PROOF. Since $p'', p' > x^{1/7}$, let $K = p'p''$. It is easy to see that

$$c_{2Kx} = \frac{p'p''}{(p' - 2)(p'' - 2)} c_{2x} = c_{2x}(1 + O(x^{-1/7})),$$

and so, by Lemma 4,

$$\sum_{\substack{p',p''>x^{1/7} \\ p'p''\leqslant x^{1/2} \\ (p'p'',x)=1}} P_{\omega_{p'p''}}\left(\frac{x}{p'p''},\, x^{1/7},\, \left(\frac{x}{p'p''}\right)^{1/2}\right)$$

$$=\sum_{\substack{p',p''>x^{1/7} \\ p'p''\leqslant x^{1/2} \\ (p'p'',x)=1}} P_{\omega_{p'p''}}\left(\frac{x}{p'p''},\, \left(\frac{x}{p'p''}\right)^{\log x/(7\log x-7\log p'p'')},\, \left(\frac{x}{p'p''}\right)^{1/2}\right)$$

$$\leqslant c_{2x}\left(1+O(x^{-1/7})\right)\sum_{\substack{p',p''>x^{1/7} \\ p'p''\leqslant x^{1/2}}}\Lambda\left(\frac{7\log x-7\log p'p''}{\log x}\right)\frac{x}{p'p''}\log^{-2}\frac{x}{p'p''}$$

$$+O\left(\frac{c_{2x}x}{\log^2 x\log\log x}\right)$$

$$=49\frac{c_{2x}x}{\log^2 x}\sum_{\substack{p',p''>x^{1/7} \\ p'p''\leqslant x^{1/2}}}\frac{\Lambda((7\log x-7\log p'p'')/\log x)}{((7\log x-7\log p'p'')/\log x)^2}\frac{1}{p'p''}$$

$$+O\left(\frac{c_{2x}x}{\log^2 x\log\log x}\right).$$

Since $\Lambda(d)/d^2$ is a decreasing function of d, we have

$$\sum_{\substack{p',p''>x^{1/7} \\ p'p''\leqslant x^{1/2}}}\frac{\Lambda((7\log x-7\log p'p'')/\log x)}{((7\log x-7\log p'p'')/\log x)^2}\frac{1}{p'p''}$$

$$\leqslant\sum_{n=20}^{34}\frac{\Lambda(6.9-0.1n)}{(6.9-0.1n)^2}\sum_{\substack{p',p''>x^{1/7} \\ x^{0.1n/7}<p'p''\leqslant x^{0.1n+0.1/7}}}\frac{1}{p'p''}$$

$$=\frac{\Lambda(3.5)}{3.5^2}\sum_{\substack{p',p''>x^{1/7} \\ p'p''\leqslant x^{1/2}}}\frac{1}{p'p''}$$

$$-\sum_{n=21}^{34}\left[\frac{\Lambda(6.9-0.1n)}{(6.9-0.1n)^2}-\frac{\Lambda(7-0.1n)}{(7-0.1n)^2}\right]\sum_{\substack{p',p''>x^{1/7} \\ p'p''\leqslant x^{0.1n/7}}}\frac{1}{p'p''}.$$

From Lemma 4 and Lemma 6, together with computation, we obtain Lemma 7.

PROOF OF THE THEOREM. From Lemma 1, Lemma 5 and Lemma 7 we get

$$\sum_{p\in J(1,3)}a_p>0.517\frac{c_{2x}x}{\log^2 x}.$$

This proves the theorem.

BIBLIOGRAPHY

1. M. B. Barban, *The "density" of the zeros of Dirichlet L-series and the problem of the sum of primes and "near primes"*, Mat. Sb. **61(103)** (1963), 418–425. (Russian)

2. Wang Yuan, *On the representation of large integer as a sum of a prime and an almost prime*, Sci. Sinica **11** (1962), 1033–1054.*

3. Pan Cheng-dong, *On the representation of even numbers as the sum of a prime and a product of not more than 4 primes*, Shantung Univ. J. **1962**, no. 2, 40–62; Russian transl., Scientia Sinica **12** (1963), 455–473.

4. Wang Yuan, *On sieve methods and some of their applications*, Acta Math. Sinica **8** (1958), 413–429; English transl., Scientia Sinica **8** (1959), 357–381; reprinted in Chinese Math. Acta **9** (1967), 121–145.

Translated by H. S. SUN

Editor's note.* Chinese version, Acta Math. Sinica **10 (1960), 168–181; English transl. in Chinese Math. Acta **1** (1962).

Amer. Math. Soc. Transl.
(2) Vol. **119**, 1983

The Indeterminate Equations
$$x^6 + y^6 = az^2, \; x^6 + y^6 = az^3, \; x^4 + y^4 = az^4*$$

V. A. DEM'JANENKO

It is well known that the genus of a curve is an invariant which to a large extent determines the structure of its set of rational points.

Curves of genus 0 and 1 have been rather thoroughly studied. For example, Hilbert and Hurwitz [1] showed that every curve of genus 0 is isomorphic to a second order plane curve $ax^2 + by^2 + cz^2 = 0$, while Mordell [2] proved that the rational points of a curve of genus 1 form a commutative group with finitely many generators. Consequently, the set of rational points on a curve of genus 0 can be effectively determined. The question of effectively determining basis points on a curve of genus 1 has not yet been completely solved.

Substantially fewer results have been obtained for curves of genus $g > 1$. As early as 1922, Mordell made the following conjecture for such curves: the number of rational points on any curve of genus $g > 1$ is finite. However, this conjecture has not yet been proved.

In this paper we investigate the following curves over the field of rational numbers:
$$x^6 + 1 = ay^2, \qquad x^6 + 1 = ay^3, \qquad x^4 + y^4 = a,$$
where we assume that the ranks of the genus one curves $u^3 + a^3 = v^2$, $u^3 - a^2 = v^2$ and $a - u^4 = v^2$ over this field do not exceed 1.

§1. Rational points of the curves $x^6 + 1 = ay^2$ and $x^6 + 1 = ay^3$

We first introduce the following notation: P is a point on the curve
$$x^6 + 1 = ay^2; \tag{1}$$

Q is a point on the curve
$$u^3 + a^3 = v^2; \tag{2}$$

1980 *Mathematics Subject Classification.* Primary 10B15, 14H45.

* Translation of Izv. Vysš. Učebn. Zaved. Matematika **1968**, no. 4 (71), 26–32; MR **37** #4007.

S is the group of rational points on the curve (2); $2S$ is the group of twice rational points on the curve (2); $T_1, \ldots, T_n$ are the elements of the quotient group $S/2S$; and O is a point of finite order on the curve (2).

We agree not to distinguish between the points $Q_1 = \{u_1, v_1\}$ and $Q_2 = \{u_1, -v_1\}$ and to take a squarefree in equations (1) and (2). Further, we see by an immediate verification that the curve (2) is the image of the curve (1) under the maps

$$\varphi_1 = \{ax^2, a^2y\}, \qquad \varphi_2 = \left\{\frac{a}{x^2}, \frac{a^2y}{x^3}\right\}, \qquad \varphi_3 = \left\{\frac{a(2x^2 - 1)}{x^2 + 1}, \frac{3a^2xy}{(x^2 + 1)^2}\right\},$$

$$\varphi_4 = \left\{\frac{a(2 - x^2)}{x^2 + 1}, \frac{3a^2y}{(x^2 + 1)^2}\right\},$$

Consequently, the maps $\varphi_i = \{u_i, v_i\}$ $(i = \overline{1,4})$ assign to each point P of the curve (1) the points $Q_i = \varphi_i(P)$ of the curve (2). Since the curves $x^6 + 1 = y^2$ and $x^6 + 1 = 2y^2$ have been completely investigated over the field $R(1)$ (see [3]), we may assume that $a \neq 1, 2$. It is not hard to show that, for the values of a satisfying these conditions, the images $Q_1, \ldots, Q_4$ of the point P are distinct.

We now prove several lemmas.

LEMMA 1. *None of the points $Q_1, \ldots, Q_4$ belong to the class $2S$.*

PROOF. We first note that the following results can be obtained using chords and tangents to the curve (2):

1) If $Q = \{u_1, v_1\}$, then $2Q = \{u_2, v_2\}$, where

$$u_2 = (u_1^4 - 8a^3u_1)/4v_1^2, \qquad v_2 = (u_1^6 + 20a^3u_1^3 - 8a^3)/8v_1^3. \tag{3}$$

2) If $Q_1 = \{u_1, v_1\}, Q_2 = \{u_2, v_2\}$, then $Q_1 \pm Q_2 = \{\eta u_\pm, v_\pm\}$, where

$$u_\pm = \frac{\mp 2v_1v_2 + u_1u_2(u_1 + u_2) + 2a^3}{(u_1 - u_2)^2},$$

$$v_\pm = \frac{v_1(u_2^3 - 3u_2^2u_1 - 2a^3) \mp v_2(u_1^3 - 3u_1^2u_2 - 2a^3)}{(u_1 - u_2)^3}. \tag{4}$$

Obviously, to prove the lemma it suffices to show that the following equations cannot be solved in rational numbers:

$$x^2 = \frac{u^4 - 8a^3u}{4av^2}, \qquad \frac{2x^2 - 1}{x^2 + 1} = \frac{u^4 - 8a^3u}{4av^2},$$

where $a \neq 1, 2$ and $x^6 + 1 = ay^2$. These equations are equivalent to the following equations, respectively:

$$mn(m^3 + n^3)(m^3 - 8n^3) = r^2 \tag{5}$$

and

$$(m^4 + 4m^3n - 8mn^3 + 4n^4)(m^4 - 8m^3n - 8mn^3 - 8n^4) = -r^2. \tag{6}$$

We first consider equation (5). Without loss of generality we may assume that $m \not\equiv 0(\mathrm{mod}\,2)$. In fact, if $m \equiv 0(\mathrm{mod}\,2)$, then, by the condition $(m, n) = 1$, we have $n \not\equiv 0(\mathrm{mod}\,2)$. Setting $m = -2n_1$, $n = m_1$, $r = 4r_1$, we obtain $m_1 n_1(m_1^3 + n_1^3)(m_1^3 - 8n_1^3) = r_1^2$, where we now have $m_1 \not\equiv 0(\mathrm{mod}\,2)$. Further, the equations $(m^3 + n^3, m^3 - 8n^3) = (1, 9)$ and (5) imply

$$p^6 + q^6 = s^2 \tag{7}$$

or

$$p^6 - q^6 = s^2. \tag{8}$$

Since equations (7), (8) only have the following rational solutions:

$$\{p, q, s\} = \{\pm 1, 0, \pm 1\}, \{0, \pm 1, \pm 1\}, \{\pm 1, \pm 1, 0\}$$

and $x^6 + 1 = ay^2$, it follows that $a = 1$. But this contradicts the assumptions of the lemma.

We now consider equation (6). If $m \not\equiv 0(\mathrm{mod}\,2)$, then, passing from (6) to congruence $\mathrm{mod}\,4$, we obtain $z^2 + 1 \equiv 0(\mathrm{mod}\,4)$, which is impossible; but if $m \equiv 0(\mathrm{mod}\,2)$, then $n \not\equiv 0(\mathrm{mod}\,2)$ and $m^4 + 4m^3n - 8mn^3 + 4n^4 \equiv 4(\mathrm{mod}\,16)$, $m^4 - 8m^3n - 8mn^3 - 8n^4 \equiv 8(\mathrm{mod}\,16)$, so that $r^2 \equiv 32(\mathrm{mod}\,64)$, which is again impossible. Thus, the lemma is proved.

LEMMA 2. *No two points among $Q_1, \ldots, Q_4$ belong to the same class T_j, $1 \leqslant j \leqslant n$.*

PROOF. In view of the analogy, it suffices to consider only two cases.

1) $Q_1, Q_2 \in T_j$. Since $Q_1 \equiv Q_2(\mathrm{mod}\,2)$, there exist rational points Q_3 and Q_4 such that $Q_1 = Q_3 + Q_4$ and $Q_2 = Q_3 - Q_4$. Consequently, by (4),

$$ax^2 = \frac{-2v_3v_4 + u_3u_4(u_3 + u_4) + 2a^3}{(u_3 - u_4)^2},$$

$$\frac{a}{x^2} = \frac{2v_3v_4 + u_3u_4(u_3 + u_4) + 2a^3}{(u_3 - u_4)^2}, \tag{9}$$

where $Q_3 = \{u_3, v_3\}$, $Q_4 = \{u_4, v_4\}$. If $u_3 = u_4$, then $v_3 = \pm v_4$, and (9) and (1) give $a = 1$, which contradicts the assumption. Multiplying the equalities (9) and canceling the $(u_3 - u_4)^2$, we obtain

$$\left[u_3(u_4 - a) - u_4 a\right]\left[u_3(u_4 + a) + u_4 a + 4a^2\right] = 0,$$

so that

$$u_3 = \frac{au_4}{u_4 - a} \quad \text{or} \quad u_3 = -\frac{au_4 + 4a^2}{u_4 + a}.$$

a) $u_3 = au_4/(u_4 - a)$. Since $u_3^3 + a^3 = v_3^2$ and $u_4^3 + a^3 = v_4^2$, we have

$$x^4 - x^2y^2 + y^4 = z^2, \tag{10}$$

where

$$x = v_3 v_4 (u_4 - a)^2, \quad y = a^2(2u_4 - a)(u_4^2 - au_4 + a^2),$$

$$z = a^3(2u_4 - a)(u_4^2 - au_4 + a^2)^3.$$

Using the method of infinite descent, we show that (10) only has the following rational solutions:

$$\{x, y, z\} = \{0,0,0\}, \{0, \pm 1, \pm 1\}, \{\pm 1, 0, \pm 1\}, \{\pm 1, \pm 1, \pm 1\}.$$

Thus, u_4 must satisfy one of the following equations:

$$(2u_4 - a)(u_4^2 - au_4 + a^2) = 0, \qquad v_3 v_4(u_4 - a) = 0,$$

$$\pm a(2u_4 - a)^2(u_4^2 - au_4 + a^2)^2 = (2u_4 - a)(u_4^2 - au_4 + a^2)^3,$$

so that $\{u_3, u_4\} = \{-a, a/2\}$, $\{a/2, -a\}$, $\{\infty, a\}$, $\{a, \infty\}$ or $\{2a, 2a\}$. Taking into account that $u_3^3 + a^3 = v_3^2$ and $u_4^3 + a^3 = v_4^2$, we find that $a = 1$ or 2, which contradicts the assumption.

b) $u_3 = -a(u_4 + 4a)/(u_4 + a)$. Substituting the value of u_3 in the equation $u_3^3 + a^3 = v_3^2$, we obtain

$$-\frac{9a^4(u_4^2 + 5au_4 + 7a^2)}{(u_4 + a)^3} = v_3^2.$$

Thus, the system

$$u_4^3 + a^3 = v_4^2, \qquad -\frac{9a^4(u_4^2 + 5au_4 + 7a^2)}{(u_4 + a)^3} = v_3^2 \qquad (11)$$

must have solutions in rational numbers. However, the system (11) is clearly not solvable, since the right sides of its equations have the same sign and the left sides have opposite signs.

2) $Q_1, Q_3 \in T_j$. As in the first case we examine the equations

$$ax^2 = \frac{-2v_3 v_4 + u_3 u_4(u_3 + u_4) + 2a^3}{(u_3 - u_4)^2},$$

$$a\frac{2x^2 - 1}{x^2 + 1} = \frac{2v_3 v_4 + u_3 u_4(u_3 + u_4) + 2a^3}{(u_3 - u_4)^2}. \qquad (12)$$

Eliminating x, we obtain $(u_3 - u_4)^4(m^2 - 432n^2) = 0$, where

$$m = u_3^2 u_4^2 - au_3 u_4(u_3 + u_4) + a^2(u_3 + u_4)^2$$
$$- 22a^2 u_3 u_4 - 22a^3(u_3 + u_4) - 20a^4,$$
$$n = a^2(u_3 u_4 + au_3 + au_4 - a^2).$$

Since $u_3 \neq u_4$, we have $m^2 - 432n^2 = 0$; m and n are rational numbers, so that $m = n = 0$, i.e.,

$$\begin{cases} u_3^2 u_4^2 - au_3 u_4(u_3 + u_4) + a^2(u_3 + u_4)^2 \\ \qquad - 22a^2 u_3 u_4 - 22a^3(u_3 + u_4) - 20a^4 = 0, \\ u_3 u_4 + a(u_3 + u_4) - a^2 = 0, \end{cases}$$

whence

$$\begin{cases} u_3^2 u_4^2 - au_3 u_4(u_3 + u_4) + a^2(u_3 + u_4)^2 = 42a^4, \\ u_3 u_4 + a(u_3 + u_4) = a^2. \end{cases}$$

But the system $x^2 - xy + y^2 = 42$, $x + y = 1$ is not solvable in rational numbers. Thus, the second case is also impossible. The lemma is proved.

THEOREM 1. *If the rank of the elliptic curve* (2) *over the field of rational numbers does not exceed* 1, *then the curve* (1) *does not have rational points, except in two cases:* $a = 1$, $\{x, y\} = \{0, \pm 1\}$; $a = 2$, $\{x, y\} = \{\pm 1, \pm 1\}$.

PROOF. We let Q denote a basis point for the curve (2). Since, by assumption, the points $P_1 = \{x, y\}$ and $P_2 = \{x, -y\}$ coincide, it follows by results of Billing [4] that the curve (2) has a unique point of finite order $O = \{-a, 0\}$. We now suppose that the curve (1) has a rational point P different from the points in the statement of the theorem. In this case $Q_i = \varphi_i(P) = a_i Q + b_i O (i = \overline{1,4})$. It is not hard to show that $Q_i - b_i O = Q_j$ where $j \in (1, 2, 3, 4)$. Hence, by Lemma 1, $a_i \not\equiv 0 \pmod 2$ $(i = \overline{1,4})$. Further, the a_i cannot coincide for all values of i. In fact, otherwise, taking into account that O is a point of order two, we would have only two distinct images for P among the Q_i, which contradicts the assumption. Now let $Q_k = a_k Q + b_k O$, $Q_r = a_r Q + b_r O$, $a_k \neq a_r$, $k, r \in (1, 2, 3, 4)$; then $a_k Q = Q_k - b_k O = Q_m$ and $a_r Q = Q_r - b_r O = Q_n$, $m, n \in (1, 2, 3, 4)$, and, since $a_k \neq a_r$ and $a_k \equiv a_r \pmod 2$, we have $Q_m \neq Q_n$ and $Q_m \equiv Q_n \pmod 2$, which contradicts Lemma 2. Thus, the curve (1) cannot have any rational points other than $\{0, \pm 1\}$ and $\{\pm 1, \pm 1\}$, Q.E.D.

We similarly prove

THEOREM 2. *If the rank of the elliptic curve* $u^3 - a^2 = v^2$ *over the field of rational numbers does not exceed* 1, *then the curve* $x^6 + 1 = ay^3$ *does not have any rational points other than* $a = 1$, $\{x, y\} = \{0, 1\}$ *and* $a = 2$, $\{x, y\} = \{\pm 1, 1\}$.

§2. Rational points of the curve $x^4 + y^4 = a$

We agree not to distinguish between the points $P = \{x, y, z\}$ and $Q = \{\pm kx, \pm ky, \pm k^2 z\}$. In this case the equation $ax^4 + by^4 = cz^2$ defines an algebraic curve. We consider the following curves with rational points:

$$a_k x_k^4 + b_k y_k^4 = z_k^2, \qquad a_k b_k = -a, \tag{13}$$

and

$$c_j u_j^4 + d_j v_j^4 = w_j^2, \qquad c_j d_j = 4a. \tag{14}$$

We easily verify that, if the points $P_1 = \{x_1, y_1, z_1\}$ and $P_2 = \{x_2, y_2, z_2\}$ belong to the curves $a_1 x_1^4 + b_1 y_1^4 = z_1^2$ and $a_2 x_2^4 + b_2 y_2^4 = z_2^2$, $a_1 b_1 = a_2 b_2$, respectively, then the point $P_1 + P_2 = \{x_3, y_3, z_3\}$ lies on the curve $a_1 a_2 x_3^4 + (b_1/a_2)y_3^4 = z_3^2$, where

$$x_3 = x_1 y_1 z_2 - x_2 y_2 z_1, \qquad y_3 = a_1 x_1^2 y_2^2 - a_2 x_2^2 y_1^2,$$
$$z_3 = z_1 z_2 (a_1 x_1^2 y_2^2 + a_2 x_2^2 y_1^2) - 2a_1 x_1 y_1 x_2 y_2 (a_2 x_1^2 x_2^2 + b_1 y_1^2 y_2^2).$$

We further agree not to distinguish between the curves $ax^4 + by^4 = cz^2$ and $aa_1^4 x_1^4 + bb_1^4 y_1^4 = cc_1^2 z_1^2$, since they are transformed into one another by means of the simple substitution $x \rightleftarrows a_1 x_1$, $y \rightleftarrows b_1 y_1$, $z \rightleftarrows c_1 z_1$. Now it is not hard to show that the curves (13) and (14) form finite groups G and U, which can be represented as a direct product of cyclic groups of order two. The units of these groups are $E_1 \colon x^4 - ay^4 = z^2$; $E_2 \colon u^4 + 4av^4 = w^2$. The sets of rational points of the curves of the groups G and U are also groups, which we denote by S and T. We further designate the groups of rational points of the curves E_1 and E_2 by S_1 and T_1. Then the quotient groups S/S_1 and T/T_1 are isomorphic to the groups G and U, respectively.

Using the formulas for adding points on a curve, we can prove the following assertions:

1) If P is a point on any of the curves (13) which is rational over a given field K, then $(1 + i)P$ is a point of the curve E_2 which is rational over the same field.

2) If Q is a point on any of the curves (14) which is rational over a given field K, then $(1 - i)Q$ is a point of the curve E_1 which is rational over the same field.

Let the groups G and U have bases consisting of the curves $M_1, \ldots, M_m$ and $N_1, \ldots, N_n$. Then the bases of the isomorphic quotient groups S/S_1 and T/T_1 consist of the cosets $S_1 + P_j (j = 1, \ldots, m)$ and $T_1 + Q_s (s = 1, \ldots, n)$. Since, by assumption, the points $P = \{x, y, z\}$ and $Q = \{\pm kx, \pm ky, \pm k^2 z\}$ are not considered to be distinct, it follows by results of Billing [4] that, for $a \neq \pm b^2$, $b \in R(1)$, the curves of the groups G and U do not have points of finite order over the field $R(1)$. Further, using the method of Podsypanin [3], we can prove the following lemmas.

LEMMA 3. *Every rational point of the curve* $\prod_{j=1}^{m} M_j^{k_j}$ *with* $a \neq \pm b^2$, $b \in R(1)$ *has the form*

$$\sum_{j=1}^{m} a_j h_j + (1 + i) \sum_{s=1}^{n} b_s g_s,$$

where $a_j \equiv k_j \pmod 2$, h_j, g_s *are bases of the groups* S *and* T ($j = 1, \ldots, m$; $s = 1, \ldots, n$).

LEMMA 4. *Every rational point of the curve $\prod_{s=1}^{n} N_s^{k_s}$ with $a \neq \pm b^2$, $b \in R(1)$ has the form*

$$(1 - i) \sum_{j=1}^{m} a_j h_j + \sum_{s=1}^{n} b_s g_s,$$

where $b_s \equiv k_s(\mathrm{mod}\, 2)$, h_j, g_s are bases of the groups S and $T(j = 1, 2, \ldots, m;$ $s = 1, 2, \ldots, n)$.

LEMMA 5. *The rank of each of the curves of the groups G and U is equal to the sum of the ranks of these groups.*

THEOREM 3. *If U is the unit group, then the curve*

$$x^4 + y^4 = a, \tag{15}$$

does not have any points in the field $R(1)$ except for the cases $a = 1, \{x, y\} = \{\pm 1, 0\}, \{0, \pm 1\}$ and $a = 2, \{x, y\} = \{\pm 1, \pm 1\}$.

PROOF. To prove the theorem it obviously suffices to consider only the case $a > 0$, $a \neq b^2$, $2b^2$. If we take any point $P = \{x_0/z_0, y_0/z_0\}$ on the curve (15), it generates two points $Q_1 = \{x_1, y_1, z_1\} = \{x_0, z_0, y_0^2\}$ and $Q_2 = \{x_2, y_2, z_2\} = \{y_0, z_0, x_0^2\}$ on the curve $ay^4 - x^4 = z^2$ such that for $a \neq \pm 1, \pm 2$

$$Q_1 \pm Q_2 \neq O, \tag{16}$$

where O is the zero of the additive group S. Since U is the unit group, it follows by Lemma 3 that these points can be represented in the form

$$Q_1 = \sum_{j=1}^{m} a_j h_j, \qquad Q_2 = \sum_{j=1}^{m} b_j h_j,$$

where $a_j \equiv b_j(\mathrm{mod}\, 2)$ $(j = 1, \ldots, m)$. Consequently, there exist rational points $P_3 = (P_1 + P_2)/2$ and $P_4 = (P_1 - P_2)/2$, which belong to the curves

$$\prod_{j=1}^{m} M_j^{(a_j + b_j)/2}, \qquad \prod_{j=1}^{m} M_j^{(a_j - b_j)/2},$$

respectively, where these curves can be written in the form

$$bx^4 - cy^4 = z^2, \qquad -bx^4 + cy^4 = z^2, \qquad bc = a.$$

The formulas for adding and subtracting points give

$$\pm a_1 x_1 = x_3 y_3 z_4 - x_4 y_4 z_3, \qquad \pm d_1 y_1 = x_3^2 y_4^2 + x_4^2 y_3^2,$$
$$\pm a_1^2 z_1 = z_3 z_4 \left(x_3^2 y_4^2 - x_4^2 y_3^2 \right) + 2b x_3 x_4 y_3 y_4 \left(c y_3^2 y_4^2 - b x_3^2 x_4^2 \right),$$
$$\pm d_2 x_2 = x_3 y_3 z_4 + x_4 y_4 z_3, \qquad \pm d_2 y_2 = x_3^2 y_4^2 + x_4^2 y_3^2,$$
$$\pm d_2^2 z_2 = z_3 z_4 \left(x_3^2 y_4^2 - x_4^2 y_3^2 \right) - 2b x_3 x_4 y_3 y_4 \left(c y_3^2 y_4^2 - b x_3^2 x_4^2 \right),$$

where $P_3 = \{x_3, y_3, z_3\}$, $P_4 = \{x_4, y_4, z_4\}$. By (16) we have $x_3 y_3 z_3$, $x_4 y_4 z_4 \neq 0$. Further, we easily see that $z_1 = x_2^2$ and $z_2 = x_1^2$, and hence

$$\pm \left(x_3 y_3 z_4 - x_4 y_4 z_3 \right)^2 = z_3 z_4 \left(x_3^2 y_4^2 - x_4^2 y_3^2 \right) - 2b x_3 x_4 y_3 y_4 \left(c y_3^2 y_4^2 - b x_3^2 x_4^2 \right),$$

so that

$$\pm 2x_3 x_4 y_3 y_4 = x_3^2 y_4^2 - x_4^2 y_3^2 \quad \text{or} \quad \pm z_3 z_4 = b x_3^2 x_4^2 + c y_3^2 y_4^2.$$

The first equation is obviously impossible. Squaring both sides of the second equation and solving it for b, we obtain

$$4x_3^4 x_4^4 \frac{b}{c} = \left(x_3^2 y_4^2 - x_4^2 y_3^2 \right)^2 \pm \sqrt{\left(x_3^2 y_4^2 - x_4^2 y_3^2 \right)^4 - \left(2x_3 x_4 y_3 y_4 \right)^4}.$$

The radicand can only be an exact square if $x_3^2 y_4^2 - x_4^2 y_3^2 = \pm 2x_3 x_4 y_3 y_4$ or $2x_3 x_4 y_3 y_4 = 0$, which is again impossible. The theorem is proved.

As a corollary, Theorem 3 yields

THEOREM 4. *If the rank of the curve E_1 over the field $R(1)$ does not exceed 1, then the curve* (15) *does not have any rational points except for the cases $a = 1$, $\{x, y\} = \{0, \pm 1\}$, $\{\pm 1, 0\}$ and $a = 2$, $\{x, y\} = \{\pm 1, \pm 1\}$.*

In fact, if $a < 0$, or $a = b^2, 2b^2$, then, for any groups G and U, the curve (15) can only have the following rational points: $\{x, y\} = \{0, \pm 1\}$, $\{\pm 1, 0\}$, $\{\pm 1, \pm 1\}$, under the condition that $a = 1$ or $a = 2$. Hence we need only consider the case $a > 0$ and $a \neq b^2, 2b^2$. By Lemma 5, the rank of the curve E_1, which, by assumption, does not exceed one, is the sum of the ranks of the groups G and U. At the same time, if the curve (15) has a rational point, then the curve $ay^4 - x^4 = z^2$, which does not coincide with the curve E_1 for $a \neq b^2$, belongs to the group G. Consequently, the rank of the group G is no less than one, and hence U is the unit group.

Moscow   Received 10/OCT/66

BIBLIOGRAPHY

1. D. Hilbert and A. Hurwitz, *Über die Diophantischen Gleichungen vom Geschlecht Null*, Acta Math. **14** (1890/91), 217–224.

2. L. Mordell, *On the rational solutions of the indeterminate equations of the third and fourth degrees*, Proc. Cambridge Philos. Soc. **21** (1922/23), 179–192.

3. V. D. Podsypanin, *On the indeterminate equation $x^3 = y^2 + Az^6$*, Mat. Sb. **24** (**66**) (1949), 391–403. (Russian)

4. G. Billing, *Beiträge zur arithmetischen Theorie der ebenen kubischen Kurven von Geschlecht Eins*, Nova Acta Regiae Soc. Sci. Upsaliensis (4) **11** (1938), no. 1.

Translated by N. KOBLITZ

Amer. Math. Soc. Transl.
(2) Vol. **119**, 1983

Algebraic Independence
of the Values of Certain E-functions*[1]

A. A. ŠMELEV

Let $\mathbf{Z}$ be the ring of integers; let $\mathbf{Q}$, $\mathbf{C}$ and $\mathbf{A}$ be the fields of rational, complex, and algebraic numbers, respectively; if $\mathbf{K}$ is an algebraic field and $f_1,\ldots,f_s \notin \mathbf{K}$, then $\mathbf{K}[f_1,\ldots,f_s]$ is the ring of polynomials in $f_1,\ldots,f_s$.

In 1929, Siegel [1] examined the functions

$$K_\tau(z) = \sum_{n=0}^{\infty} \frac{(-1)^n}{n!\,(\tau + 1)\cdots(\tau + n)}\left(\frac{z}{2}\right)^{2n}, \quad \tau \neq -1,-2,\ldots, \tag{1}$$

which satisfy the differential equation

$$y'' + (2\tau + 1)y'/z + y = 0, \tag{2}$$

and proved that, if $\tau_1,\ldots,\tau_m \in \mathbf{Q}$, $\tau_i \neq 0$, $\pm\frac{1}{2}$, $-1,\ldots,(m \geqslant 1)$, $\tau_{i_1} \pm \tau_{i_2} \notin \mathbf{Z}$ $(0 \leqslant i_1 < i_2 \leqslant m)$, $\xi_1,\ldots,\xi_n \in \mathbf{A}\backslash\{0\}$ and $\xi_{j_1}^2 \neq \xi_{j_2}$ $(1 \leqslant j_1 < j_2 \leqslant n)$, then the $2mn$ numbers $K_{\tau_i}(\xi_j)$, $K'_{\tau_i}(\xi_j)$ $(i = 1,\ldots,m;\ j = 1,\ldots,n)$ are algebraically independent over $\mathbf{Q}$.

In 1954, Šidlovskiĭ [2] examined the functions

$$K_{\lambda\mu}(z) = \sum_{n=0}^{\infty} \frac{(-1)^n}{(\lambda + 1)\cdots(\lambda + n)(\mu + 1)\cdots(\mu + n)}\left(\frac{z}{2}\right)^{2n},$$
$$\lambda, \mu \neq -1,-2,\ldots, \tag{3}$$

which satisfy the differential equation

$$y'' + (2\lambda + 2\mu + 1)y'/z + \left(1 + 4\lambda\mu/z^2\right)y = 4\lambda\mu/z^2, \tag{4}$$

and proved that, if λ, $\mu \in \mathbf{Q}$, λ, $\mu \neq -1,-2,\ldots,\lambda - \mu \neq (2k + 1)/2$, $k \in \mathbf{Z}$, $\alpha \in \mathbf{A}\backslash\{0\}$, then the numbers $K_{\lambda\mu}(\alpha)$, $K'_{\lambda\mu}(\alpha)$ are algebraically independent over $\mathbf{Q}$. In 1966, Šidlovskiĭ [3] gave a significant simplification of the proof of his theorem in [2].

1980 *Mathematics Subject Classification.* Primary 10F37.
*Translation of Izv. Vysš. Učebn. Zaved. Matematika **1969**, no. 4 (83), 103–112; MR **40** #96.
[1] For the definition of E-functions, see [1], [2] or [4].

In this paper we use Šidlovskiĭ's method in [2] to prove(^{2}) the following result.

THEOREM. *Let* $\lambda_1,\ldots,\lambda_m$, $\mu_1,\ldots,\mu_m$, $\tau_1,\ldots,\tau_l \in \mathbf{Q}$, $\beta_i = \lambda_i - \mu_i$ $(i = 1,\ldots,m)$, λ_i, μ_i, $\tau_j \neq -1,-2,\ldots,\beta_j$, $\tau_j \neq (2k+1)/2$, $k \in \mathbf{Z}$ $(i = 1,\ldots,m; j = 1,\ldots,l)$, $\beta_{i_1} \pm \beta_{i_2}$, $\tau_{s_1} \pm \tau_{s_2}$, $\beta_i \pm \tau_j \notin \mathbf{Z}$ $(1 \leq i_1 < i_2 \leq m, 1 \leq s_1 < s_2 \leq l, i = 1,\ldots,m; j = 1,\ldots,l)$, $\alpha_1,\ldots,\alpha_n \in \mathbf{A}\setminus\{0\}$ *and* $\alpha_{k_1}^2 \neq \alpha_{k_2}^2$ $(1 \leq k_1 < k_2 \leq n)$.

Then the $2(l + m)n$ *numbers* $K_{\lambda_i\mu_i}(\alpha_j)$, $K'_{\lambda_i\mu_i}(\alpha_j)$, $K_{\tau_s}(\alpha_j)$, $K'_{\tau_s}(\alpha_j)$ $(i = 1,\ldots,m; j = 1,\ldots,n; s = 1,\ldots,l)$ *are algebraically independent over* $\mathbf{Q}$.

LEMMA 1 [1]. *Suppose that* $\eta_1,\ldots,\eta_p \in \mathbf{C}$, $\eta_i \neq 0$, $\pm\frac{1}{2},-1,\ldots,\eta_{i_1} \pm \eta_{i_2} \notin \mathbf{Z}$ $(1 \leq i_1 < i_2 \leq m)$, *and that* $v_\lambda(z)$ *and* $w_\lambda(z)$ *are arbitrary linearly independent solutions of the equation*

$$y'' + y'/z + (1 - \lambda^2/z^2)y = 0, \tag{5}$$

$\xi_1,\ldots,\xi_n \in \mathbf{C}$, $\xi_{l_1}^2 \neq \xi_{l_2}^2$ $(1 \leq l_1 < l_2 \leq m)$.

Then the $3pn + 1$ *functions* z, $v_{\eta_j}(\xi_l z)$, $v'_{\eta_i}(\xi_l z)$, $w_{\eta_i}(\xi_l z)$ $(i = 1,\ldots,p; j = 1,\ldots,n)$ *are algebraically independent over* $\mathbf{C}$.

If R is a rational function in $f_1(z),\ldots,f_p(z)$, which satisfies the system of differential equations

$$y'_k = Q_{k0}(z) + \sum_{i=1}^{p} Q_{ki}(z), \quad k = 1,\ldots,p, \tag{6}$$

then by R' we mean the total derivative of R with respect to z in which $f'_k(z)$ is replaced by $Q_{k0}(z) + \Sigma_{i=1}^{p} Q_{ki}(z)f_i$.

LEMMA 2. *Suppose that* $\eta_1,\ldots,\eta_p$ *and* $\xi_1,\ldots,\xi_n$ *satisfy the conditions of Lemma* 1; $v_{ij} = v_{\eta_i}(\xi_j z)$ $(i = 1,\ldots,p; j = 1,\ldots,n)$; *and*

$$Q \in \mathbf{C}[z, v_{11}, v'_{11},\ldots,v_{pn}, v'_{pn}], \qquad Q \not\equiv 0.$$

Then the polynomial z^2Q' *is divisible by the polynomial* Q *if and only if* Q *does not contain the variables* v_{ij} *or* v'_{ij} $(i = 1,\ldots,p; j = 1,\ldots,n)$ *and it has the form*

$$Q = cz^b, \qquad c \in \mathbf{C}\setminus\{0\}, b \in \mathbf{Z}, b \geq 0. \tag{7}$$

The case $p = n = 1$ is examined in [2]. The proof is similar in the general case.

Let $w_{ij} = w_{\eta_i}(\xi_j z)$ $(i = 1,\ldots,p; j = 1,\ldots,n)$. By the Liouville-Ostrogradskiĭ theorem,

$$v_{ij}w'_{ij} - w_{ij}v'_{ij} = \gamma_{ij}/z, \qquad \gamma_{ij} \in \mathbf{C}\setminus\{0\}. \tag{8}$$

LEMMA 3. *Suppose that* $\eta_1,\ldots,\eta_p$, $\xi_1,\ldots,\xi_n$ *satisfy the conditions of Lemma* 1, *and* $Q = Q(z, v_{11}, v'_{11}, w_{11},\ldots,v_{pn}, v'_{pn}, w_{pn}) \in \mathbf{C}(z, v_{ij}, v'_{ij}, w_{ij})$ $(i = 1,\ldots,p; j = 1,\ldots,n)$.

Under these conditions the polynomial $z^2(\prod_{i=1}^{p}\prod_{j=1}^{n} v_{ij})\, Q'$ is divisible by Q if and only if Q does not depend on the variables w_{ij} and has the form

$$Q = \sigma z^b \prod_{i=1}^{p} \prod_{j=1}^{n} v_{ij}^{k_{ij}}, \qquad \sigma \in \mathbf{C}\backslash\{0\}, b, k_{11}, \ldots, k_{pn} \in \mathbf{Z}, \tag{9}$$

$$b \geqslant 0,\, k_{ij} \geqslant 0 \; (i = 1,\ldots,p;\, j = 1,\ldots,n).$$

The case $p = n = 1$ is examined in [2]. In the general case the proof uses induction and is similar.

If $\beta_1,\ldots,\beta_m \in \mathbf{Q}$ we define the functions

$$u_{ij}(z) = z^{-\beta_i} \int_{\delta}^{z} t^{\beta_i - 1} v_{ij}(t)\, dt, \qquad s_{ij}(z) = z^{-\beta_i} \int_{\delta}^{z} t^{\beta_i - 1} w_{ij}(t)\, dt, \tag{10}$$

$$i = 1,\ldots,p;\, j = 1,\ldots,n;\, \delta > 0.$$

Then

$$u_{ij}'(z) = -\frac{\beta_i}{z} u_{ij} + \frac{v_{ij}}{z}, \qquad s_{ij}'(z) = -\frac{\beta_i}{z} s_{ij} + \frac{w_{ij}}{z}, \tag{11}$$

$$i = 1,\ldots,p;\, j = 1,\ldots,n.$$

LEMMA 4. *Suppose that $\eta_1,\ldots,\eta_p$ and $\xi_1,\ldots,\xi_n$ satisfy the conditions of Lemma 1, $\beta_1,\ldots,\beta_m \in \mathbf{Q}$ and $\eta_i \pm \beta_i \neq 2k,\, k \geqslant 1,\, k \in \mathbf{Z}\,(i = 1,\ldots,p).$*

Then the functions $v_{ij},\, v_{ij}',\, w_{ij},\, u_{ij},\, z\;(i = 1,\ldots,p;\, j = 1,\ldots,n)$ are algebraically independent over $\mathbf{C}$.

We first suppose that the functions $z,\, v_{ij},\, v_{ij}',\, w_{ij}\;(i = 1,\ldots,p;\, j = 1,\ldots,n)$ and any k functions from the set $u_{ij}\;(i = 1,\ldots,p;\; j = 1,\ldots,n)$ are algebraically independent; we index these functions $u_{i_1 j_1},\ldots,u_{i_k j_k}$. Let $P \not\equiv 0$ be a polynomial in these functions containing at least one of the variables $u_{i_l j_l}\;(1 \leqslant l \leqslant k)$ and such that $z^2(\prod_{i=1}^{p}\prod_{l=1}^{n} v_{ij})\, P'$ is divisible by P. We consider P as a polynomial in the variables $u_{i_1 j_1},\ldots,u_{i_k j_k}$ with coefficients in $\mathbf{C}[z, v_{11}, v_{11}', w_{11},\ldots,v_{pn}, v_{pn}', w_{pn}]$. We divide P into groups of homogeneous terms of the same degree in the $u_{i_l j_l}$ $(1 \leqslant l \leqslant k)$, and we take the terms of highest degree to be the leading terms. We shall assume that the order of $u_{i_t j_t}$ is greater than the order of $u_{i_q j_q}$ if $t > q$. We order each group of homogeneous terms lexicographically in the powers of $u_{i_l j_l}$ $(1 \leqslant l \leqslant k)$ according to decreasing orders of $u_{i_l j_l}$.

Let

$$P = Q_0 u_{i_l j_l}^{m_l} \cdots u_{i_1 j_1}^{m_1} + \cdots, \qquad Q_0 \in \mathbf{C}\!\left[z, v_{ij}, v_{ij}', w_{ij}\right], \qquad Q_0 \not\equiv 0,\, 1 \leqslant l \leqslant k,$$

$$m_l \geqslant 1,\, \sum_{i=1}^{l} m_i \geqslant 1,$$

where the leading term of P is shown. In the expansion

$$P' = \left\{ Q_0' + Q_0 \sum_{n=1}^{l} m_n \frac{u'_{i_n j_n}}{u_{i_n j_n}} \right\} u_{i_l j_l}^{m_l} \cdots u_{i_1 j_1}^{m_1} + \cdots$$

$$= \left\{ Q_0' - Q_1 \sum_{n=1}^{l} m_n \frac{\beta_{i_n}}{z} \right\} u_{i_l j_l}^{m_l} \cdots u_{i_1 j_1}^{m_1} + \cdots$$

the term is shown having the same degree as the leading term of P. The degrees of the polynomials P and $z^2 (\prod_{i=1}^{n} \prod_{j=1}^{n} v_{ij}) P'$ in all the variables $u_{i_1 j_1}, \ldots, u_{i_k j_k}$ are the same, while their quotient does not depend on $u_{i_1 j_1}, \ldots, u_{i_k j_k}$ and is equal to the quotient from dividing the coefficients of the leading terms; thus, with $\omega = \sum_{n=1}^{l} m_n \beta_{i_n}$,

$$P'/P = Q_0'/Q_0 + \omega/z, \qquad P = cQ_0 z^{-\omega}, \qquad c \in \mathbf{C} \backslash \{0\}, \tag{12}$$

which contradicts the assumption that the functions in P are algebraically independent. Therefore, P does not depend on $u_{i_l j_l}$ ($1 \leqslant l \leqslant k$), and, by Lemma 3, it has the form (9).

Suppose that the functions z, v_{ij}, v_{ij}', w_{ij}, u_{qj} ($i = 1,\ldots,p$; $j = 1,\ldots,n$; $q = 1,\ldots,l-1$, $l \geqslant 1$) and $u_{l1},\ldots,u_{lk-1}$ ($k \geqslant 1$) are algebraically independent over $\mathbf{C}$, but that, when u_{lk} is adjoined, they become algebraically dependent; $P \not\equiv 0$ is an irreducible polynomial in $d = (3p + l - 1)n + k + 1$ variables such that

$$P\big(z, v_{11}, v_{11}', \ldots, v_{pn}, v_{pn}', w_{11}, \ldots, w_{pn}, u_{11}, \ldots, u_{lk}\big) = 0.$$

It is obvious that $z^2 (\prod_{i=1}^{p} \prod_{j=1}^{n} v_{ij}) P'$ is divisible by P as a polynomial in d varibles. Let

$$P = P_s u_{lk}^s + \cdots + P_0, \qquad s \geqslant 1, \, P_s \not\equiv 0, \tag{13}$$

where the P_i are polynomials in the remaining variables. Differentiating P/P_s and adding the result to equality (13) multiplied by $s\beta_t/zP_s$, we obtain

$$R = \left[sv_{lk}/z + (d/dz)(P_{s-1}/P_s) + (\beta_l/z)(P_{s-1}/P_s) \right] u_{lk}^{s-1} + \cdots = 0.$$

R has degree less than s in u_{lk}, so that $R \equiv 0$ and

$$sv_{lk} P_s + zP_{s-1}' - \frac{zP_1'}{P_s} P_{s-1} + \beta_l P_{s-1} = 0, \tag{14}$$

where we may assume that $(P_{s-1}, P_s) = 1$ in this identity. Multiplying (14) by $z \prod_{i=1}^{p} \prod_{j=1}^{n} v_{ij}$, we find that $z^2 (\prod_{i=1}^{p} \prod_{j=1}^{n} v_{ij}) P_s'$ is divisible by P_s, and hence P_s has the form (9). Substituting (9) into (14), we obtain

$$\sigma s z^b v_{lk} \prod_{i=1}^{p} \prod_{j=1}^{n} v_{ij}^{k_{ij}} + zP_{s-1}' + \left[(\beta_l - b) - z \prod_{i=1}^{p} \prod_{j=1}^{n} k_{ij} \frac{v_{ij}'}{v_{ij}} \right] P_{s-1} = 0. \tag{15}$$

We show that P_{s-1} does not depend on w_{ij} ($i = 1,\ldots,p$; $j = 1,\ldots,n$). Suppose the contrary. We shall assume that the order of $w_{i_2 j_2}$ is greater than the order of $w_{i_1 j_1}$ if $i_2 > i_1$, and, in the case $i_2 = i_1$, if $j_2 > j_1$. We consider P_{s-1} as a polynomial in the functions w_{ij} $(i = 1,\ldots,p; j = 1,\ldots,n)$ whose coefficients are polynomials in z, v_{ij}, v'_{ij} $(i = 1,\ldots,p; j = 1,\ldots,n)$, u_{rj} $(r = 1,\ldots,l-1; j = 1,\ldots,n)$ and $u_{l1},\ldots,u_{lk-1}$. The terms of the polynomial are ordered in the above way. Let

$$P_{s-1} = Q_{m_1n_1}w_{m_sn_s}^{l_{m_sn_s}} \cdots w_{m_1n_1}^{l_{m_1n_1}}w_{m_1n_1} + \cdots, \tag{16}$$

where the leading term $l_{m_in_i}$ is shown and $w_{m_1n_1}$ is the variable of least order in the leading term. We consider the terms

$$Q_{r_1t_1}w_{m_sn_s}^{l_{m_sn_s}} \cdots w_{m_1n_1}^{l_{m_1n_1}}w_{r_1t_1}, \tag{17}$$

where the order of the variable $w_{r_1t_1}$ is no greater than the order of $w_{m_1n_1}$. From (15) we find that

$$z\left\{Q_{r_1t_1} + Q_{r_1t_1}\sum_{i=1}^{s}\frac{v'_{m_in_i}}{v_{m_in_i}}l_{m_in_i} + \frac{v'_{r_1t_1}}{v_{r_1t_1}}Q_{r_1t_1}\right\}$$

$$+ \left[(\beta_l - b) - z\sum_{i=1}^{p}\sum_{j=1}^{n}k_{ij}\frac{v'_{ij}}{v_{ij}}\right]Q_{r_1t_1} = 0. \tag{18}$$

We integrate this:

$$Q_{r_1t_1} = c_{r_1t_1}z^{b-\beta_l}\prod_{i=1}^{p}\prod_{j=1}^{n}v_{ij}^{k_{ij}-l_{ij}}v_{r_1t_1}^{-1}, \qquad c_{r_1t_1} \in \mathbf{C}. \tag{19}$$

It follows from (19) that $k_{r_1t_1} - l_{r_1t_1} - 1 \geqslant 0$. In (15) we equate the coefficients of $w_{m_sn_s}^{l_{m_sn_s}} \cdots w_{m_1n_1}^{l_{m_1n_1}}$; let Q_0 be the coefficient of this term in P_{s-1}, with

$$Q_0 = \prod_{i=1}^{p}\prod_{j=1}^{n}v_{ij}^{p_{ij}}Q_1, \qquad \left(Q_1, \prod_{i=1}^{p}\prod_{j=1}^{n}v_{ij}\right) = 1.$$

We obtain

$$\left\{(\beta_l - b) - z\prod_{i=1}^{p}\prod_{j=1}^{n}(k_{ij} - l_{ij} - p_{ij})\frac{v'_{ij}}{v_{ij}}\right\}Q_1$$

$$+ zQ'_1 + \sum_{(r_1,t_1)}\frac{\gamma_{r_1t_1}c_{r_1t_1}}{v_{r_1t_1}^{-2}}z^{b-\beta_l}\prod_{i=1}^{p}\prod_{j=1}^{n}v_{ij}^{k_{ij}-l_{ij}-p_{ij}}$$

$$+ l_{m_1n_1}\gamma_{m_1n_1}c_{m_1n_1}z^{b-\beta_l}\prod_{i=1}^{p}\prod_{j=1}^{n}v_{ij}^{k_{ij}-l_{ij}-p_{ij}} + \varepsilon\sigma z^bv_{lk}\prod_{i=1}^{p}\prod_{j=1}^{n}v_{ij}^{l_{ij}-p_{ij}} = 0, \tag{20}$$

where $\varepsilon = 1$ if $\sum_1^s l_{m_in_i} = 0$, and $\varepsilon = 0$ if $\sum_1^s l_{m_in_i} \geqslant 1$.

Let (r_0, t_0) be a vector in the set (r_1, t_1) such that $c_{r_0,t_0} \neq 0$. We take the third summand multiplied by $\prod_{i=1}^{p}\prod_{j=1}^{n}v_{ij}$ and select the term with index (r_0, t_0). It follows from (20) that for $(i, j) = (r_1, t_1)$ we have $k_{ij} - l_{ij} - p_{ij} \geqslant 1$. On the other hand, if $k_{i_1j_1} - l_{i_1j_1} - p_{i_1j_1} = 2$ for some (r_1, t_1), then it follows from (20) that Q_1 is divisible by $v_{r_1t_1}$, which contradicts the choice of Q_1. Thus, the term we selected does not depend on $v_{r_0t_0}$ or $v'_{r_0t_0}$, which contradicts the algebraic dependence of the functions in (20).

Thus, suppose P_{s-1} does not depend on w_{ij} $(i = 1,\ldots,p;\ j = 1,\ldots,n)$. It follows from (14), (11) and (12) that $z^2 P_s'$ is divisible by P_s; according to Lemma 2, $P_s = \sigma z^b$, and (20) has the form

$$\sigma s z^b v_{lk} + z P_{s-1}' + (\beta_l - b) P_{s-1} = 0. \tag{21}$$

If P_{s-1} has total degree $r_1 \geq 2$ in v_{ij}, v_{ij}', u_{rj}, and T_{r_1} is the group of terms of degree r_1, then from (21) we obtain $-z T_{r_1}' = (\beta_l - b) T_{r_1}$ and $T_{r_1} = c z^{b-\beta_l}$, $c \neq 0$, $c \in \mathbf{C}$, which contradicts the algebraic independence of these functions. Thus

$$P_{s-1} = \sum_{i,j} Q_{ij} v_{ij} + \sum_{i,j} R_{ij} v_{ij}' + \sum_{i,j} S_{ij} u_{ij} + T_0, \tag{22}$$

Q_{ij}, R_{ij}, S_{ij}, $T_0 \in \mathbf{C}|z|$. It follows from (21) that $T_0 = c_0 z^{b-\beta_l}$ and $S_{ij} = c_{ij} z^{b-\beta_l-\beta_i}$, c_0, $c_{ij} \in \mathbf{C}\backslash\{0\}$. Using the algebraic independence of the functions v_{ij}, v_{ij}', u_{rj}, z, from (22) and (21) we obtain

$$\varepsilon_{ij} s \sigma z^b v_{ij} + z \left\{ v_{ij} Q_{ij}' + v_{ij}' Q_{ij} + R_{ij}' v_{ij}' - \left[\frac{v_{ij}'}{z} + \left(\xi_j^2 - \frac{\eta_i^2}{z^2} \right) v_{ij} \right] R_{ij} \right\}$$

$$+ (b - \beta_l)(v_{ij} Q_{ij} + v_{ij}' R_{ij}) + c_{ij} \beta_i z^{b-\beta_l-\beta_i} v_{ij} = 0.$$

where $\varepsilon_{ij} = 0$ for $(i,\,j) = (l,\,k)$ and $\varepsilon_{ij} = 1$ for $(i,\,j) = (l,\,k)$, or

$$z Q_{ij}' + (b - \beta_l - 1) R_{ij} + z R_{ij}' = 0, \tag{23}$$

$$z Q_{ij}' - \left(\xi_j^2 z - \eta_i^2/z \right) R_{ij} + (b - \beta_l) Q_{ij} = \varepsilon_{ij} s \sigma z^b + c_{ij} \beta_l z^{b-\beta_l-\beta_i}. \tag{24}$$

From (23) and (24) we obtain

$$z R_{ij}' + (2\beta_l - 2b - 1) R_{ij} + \left[z \xi_j^2 + \frac{(\beta_i - b - 1)^2 - \eta_i^2}{z} \right] R_{ij}$$

$$= \varepsilon_{ij} s \sigma z^b + c_{ij} z^{b-\beta_l-\beta_i}. \tag{25}$$

In the case $(i,\,j) \neq (l,\,k)$ with $c_{ij} \neq 0$ equation (25) only has a polynomial solution in the case $\beta_i \pm \eta_i \neq 2k$ $(k = 1, 2, \ldots)$ ([2], Lemma 10). If $c_{ij} = 0$, we set

$$\theta_1 = (\beta_l - b - 1 + \eta_l)/2, \qquad \theta_2 = (\beta_l - b - 1 - \eta_l)/2.$$

Let $R_{ij}^* = R_{ij}(z/\xi_j)$. Then R_{ij}^* satisfies the equation

$$y'' + \frac{2\theta_1 + 2\theta_2 + 1}{z} y' + \left(1 + \frac{4\theta_1\theta_2}{z^2} \right) y = 0,$$

and the function $\tilde{R}_{ij} = z^{\theta_1+\theta_2} R_{ij}^*$ satisfies the equation

$$y'' + \frac{y'}{z} + \left[1 - \frac{(\theta_1 - \theta_2)^2}{z^2} \right] y = 0. \tag{26}$$

Siegel [1] showed that any solution of (26) is a transcendental function. Since all the u_{ij} have order lower than u_{lk}, it follows that in all cases when $c_{ij} \neq 0$ we have $\varepsilon_{ij} = 0$. Thus, $R_{ij} = 0$, and (24) and (25) imply that $Q_{ij} \equiv 0$, $T_0 \equiv 0$ and $\sigma = 0$. The lemma is proved.

LEMMA 5. *Let $\xi_1,\ldots,\xi_n$ and $\eta_1,\ldots,\eta_p$ satisfy the conditions of Lemma 1, and let $\beta_1,\ldots,\beta_p$ satisfy the conditions of Lemma 4.*

Then the functions z, v_{ij}, v'_{ij}, u_{ij}, w_{ij}, s_{ij} $(i=1,\ldots,p; j=1,\ldots,n)$ are algebraically independent over $\mathbf{C}$.

Suppose the contrary, namely, z, v_{ij}, v'_{ij}, w_{ij}, u_{ij} $(i=1,\ldots,p; j=1,\ldots,n)$, s_{rj} $(r=1,\ldots,l-1; j=1,\ldots,n)$, $s_{l1},\ldots,s_{lk-1}$ are algebraically independent over $\mathbf{C}$, but, when s_{lk} is adjoined, they become algebraically dependent. Suppose that $d=(4p+l-1)n+k+1$ and $P\in\mathbf{C}\,|\,x_1,\ldots,x_d|$ is an irreducible polynomial such that

$$P\big(z, v_{11}, v'_{11},\ldots,v_{pn}, v'_{pn}, w_{11},\ldots,w_{pn}, u_{11},\ldots,u_{pn}, s_{11},\ldots,s_{lk}\big)=0,$$

$$P=P_r s_{lk}^r+\cdots+P_0,$$

where the P_i are polynomials in the remaining variables that appear in P. Repeating the reasoning used to derive (14), we obtain

$$rw_{lk}P_r+zP'_{r-1}-\big(zP'_r/P_r\big)P_{r-1}+\beta_1 P_{r-1}=0, \tag{27}$$

where $(P_r, P_{r-1})=1$. Repeating the reasoning at the beginning of Lemma 4, we obtain

$$P_r=cz^b\prod_{i=1}^{p}\prod_{j=1}^{n}v_{ij}^{k_{ij}}, \qquad c\in\mathbf{C}\backslash\{0\}, k_{ij}\geqslant 0, k_{ij}\in\mathbf{Z}.$$

Substituting the expression for P_r in (27), we get

$$crw_{lk}z^b\prod_{i=1}^{p}\prod_{j=1}^{n}v_{ij}^{k_{ij}}+zP'_{r-1}+\left(\beta_l-b-z\sum_{i,j}k_{ij}\frac{v'_{ij}}{v_{ij}}\right)P_{r-1}=0. \tag{28}$$

Further, repeating the corresponding arguments from Lemma 4, we find that P_{r-1} does not depend on s_{ij} and has the form

$$P_{r-1}=\sum_{i,j}Q_{ij}w_{ij}+Q_0, \tag{29}$$

where the Q_{ij} are polynomials in v_{ij}, v'_{ij}, u_{ij} $(i=1,\ldots,p; j=1,\ldots,n)$ and z. It follows from (29) that

$$P'_{r-1}=\sum_{i,j}\left(Q'_{ij}+\frac{v'_{ij}}{v_{ij}}Q_{ij}\right)w_{ij}+\frac{\gamma_{ij}}{zv_{ij}}Q_{ij}+Q'_0. \tag{30}$$

From (28)–(30) we obtain

$$\varepsilon_{i_1j_1}rcz^b\prod_{i=1}^{p}\prod_{j=1}^{n}v_{ij}^{k_{ij}}+z\left(Q_{i_1j_1}+\frac{v'_{i_1j_1}}{v_{i_1j_1}}Q_{i_1j_1}\right)$$

$$+\left(\beta_l-b-z\sum_{i,j}k_{ij}\frac{v'_{ij}}{v_{ij}}\right)Q_{i_1j_1}=0, \tag{31}$$

where $\varepsilon_{i_1 j_1} = 0$ for $(i_1, j_1) \neq (l, k)$, $\varepsilon_{lk} = 1$ and

$$\sum_{i,j} \frac{\gamma_{ij}}{v_{ij}} Q_{ij} + z Q_0' + \left(\beta_l - b - z \sum_{i,j} k_{ij} \frac{v_{ij}'}{v_{ij}} \right) Q_0 = 0. \tag{32}$$

1. Suppose that at least one of the numbers k_{ij} is ≥ 2; then all the polynomials $Q_{i_1 j_1}$ are divisible by v_{ij}. It follows from (28) that P_{r-1} is divisible by v_{ij} and that $(P_r, P_{r-1}) \neq 1$, contrary to the assumption.

2. Suppose that $k_{lk} = 0$ and $k_{ij} \geq 0$, $(i, j) \neq (l, k)$; it follows from (32) that Q_{lk} is divisible by v_{lk}, $Q_{lk} = v_{lk} Q_1$. Then

$$rcz^b \prod_{i,j} v_{ij}^{k_{ij}} + z v_{lk} Q_1' + z v_{lk}' Q_1 + \left[(\beta_l - b) v_{lk} - \sum_{i,j} k_{ij} \frac{v_{ij}'}{v_{ij}} v_{lk} \right] Q_1 = 0.$$

This relation is a contradiction, since all the terms except for the first depend on v_{lk} and v_{lk}'.

3. $k_{lk} = 1$. Then

$$rcz^b \prod_{i,j} v_{ij}^{k_{ij}} + z Q_{lk}' - z \sum_{(i,j) \neq (l,k)} k_{ij} \frac{v_{ij}'}{v_{ij}} Q_{lk} + (\beta_l - b) Q_{lk} = 0,$$

which implies that

$$Q_{lk} = \sigma z^{b - \beta_l} \prod_{(i,j) \neq (l,k)} v_{ij}^{k_{ij}} - \frac{rcu_{lk}}{\xi_k} z^b \prod_{(i,j) \neq (l,k)} v_{ij}^{k_{ij}}, \qquad \sigma \in \mathbf{C} \setminus \{0\}.$$

If $(i_1, j_1) \neq (l, k)$, then from (31) we obtain

$$Q_{i_1 j_1} = c_{i_1 j_1} z^{b - \beta_l} \prod_{i,j} v_{ij}^{k_{ij}} v_{i_1 j_1}^{-1}, \qquad c_{i_1 j_1} \in \mathbf{C} \setminus \{0\}.$$

We substitute the resulting expressions in (32):

$$z^{b - \beta_l} \prod_{i,j} v_{ij}^{k_{ij}} \left(\sum_{(i_1 j_1) \neq (l,k)} \frac{c_{i_1 j_1} \gamma_{i_1 j_1}}{v_{i_1 j_1}^2} \right) + \frac{\gamma_{lk}}{v_{lk}} \prod_{(i,j) \neq (l,k)} v_{ij}^{k_{ij}} + \gamma_{lk} rcu_{lk} z^b \prod_{(i_1, j_1) \neq (l,k)} v_{ij}^{k_{ij}}$$

$$= z Q_0' + \left(\beta_l - b - z \sum_{i,j} k_{ij} \frac{v_{ij}'}{v_{ij}} \right) Q_0. \tag{33}$$

If we multiply both sides of this equality by $z \prod_{i=1}^{p} \prod_{j=1}^{n} v_{ij}$, then they become polynomials in z, v_{ij}, v_{ij}', u_{ij} $(i = 1, \ldots, p; \ j = 1, \ldots, n)$. It follows from (5) and (11) that, when we differentiate Q_0, the set of homogeneous terms of any degree in u_{lk}, v_{lk}, v_{lk}' becomes a set of similar terms. The left side of (33) contains a term with u_{lk} and without v_{lk} or v_{lk}', while the right side has no terms which do not contain either v_{lk} or v_{lk}', which contradicts the algebraic independence of the functions in (33).

PROOF OF THE THEOREM. From Šidlovskiǐ's main theorem [4] it follows that the numbers in the conditions of the theorem are algebraically independent if and only if the $2(l + m)n + 1$ functions z, $K_{\lambda_i \mu_i}(\alpha_j z)$, $K_{\lambda_i \mu_i}'(\alpha_j z)$, $K_{\tau_s}(\alpha_j z)$, $K_{\tau_s}'(\alpha_j z)$ are algebraically independent over $\mathbf{C}$. Let y_p be any solution of (4), where λ and μ

are replaced by λ_p and μ_p, respectively, and $x_p^* = z^{\lambda_p + \mu_p} y_p$. The function x_p^* is a solution of the equation

$$y'' + y'/z + \left[1 - (\lambda_p - \mu_p)^2/z^2\right] y = 4\lambda_p \mu_p z^{\lambda_p + \mu_p - 2}. \tag{34}$$

Let v_p and w_p be linearly independent solutions of (5), where λ is replaced by $\lambda_p - \mu_p$, and let φ_p be an arbitrary solution of (34). Applying the method of variation of parameters, we easily find that

$$\varphi_{pj}(z) = \varphi_p(\alpha_j z)$$
$$= (\alpha_j z)^{-(\lambda_p + \mu_p)} \left\{ c_{pj} v_p(\alpha_j z) + d_{pj} w_p(\alpha_j z) + l_{pj}\big(u_{pj} w_p(\alpha_j z) - s_{pj} v_p(\alpha_j z)\big)\right\}, \tag{35}$$

where the c_{pj}, d_{pj} and l_{pj} are constants, with $l_{pj} \neq 0$ if $\lambda_p, \mu_p \neq 0$, $\lambda_p, \mu_p \notin \mathbf{Z}$, and

$$\left. \begin{aligned}
u_{pj}(z) &= z^{-(\lambda_p + \mu_p)} \int_\delta^z v_p(\alpha_j \theta) \theta^{(\lambda_p + \mu_p) - 1} \, d\theta, \\
s_{pj}(z) &= z^{-(\lambda_p + \mu_p)} \int_\delta^z w_p(\alpha_j \theta) \theta^{(\lambda_p + \mu_p) - 1} \, d\theta, \\
&\qquad \delta > 0.
\end{aligned} \right\}$$

We suppose that the functions z, $K_{\lambda_i \mu_i}(\alpha_j z)$, $K'_{\lambda_i \mu_i}(\alpha_j z)$, $K_{\tau_2}(\alpha_j z)$, $K'_{\tau_s}(\alpha_j z)$ $(i = 1, \ldots, l;\ s = 1, \ldots, m;\ j = 1, \ldots, n)$ are algebraically independent over $\mathbf{C}$, and $P \in \mathbf{C}[x_1, \ldots, x_{2(l+m)n+1}]$, $P \not\equiv 0$, is a polynomial irreducible over $\mathbf{C}$ such that

$$P\left[z,\ K_{\lambda_1 \mu_1}(\alpha_1 z), \ldots, K'_{\tau_m}(\alpha_n z)\right] = 0. \tag{36}$$

Suppose that $\mu_i \in \mathbf{Z}$, where if $\lambda_i, \mu_i \geqslant 0$, $\lambda_i, \mu_i \in \mathbf{Z}$, then μ_i is the least one of them. Then

$$K_{\lambda_i \mu_i}(z) \frac{(-1)^{\mu_i}}{\mu_i!\,(\lambda_i - \mu_i + 1) \cdots \lambda_i} \left(\frac{z}{2}\right)^{2\mu_i}$$
$$+ \sum_{n=0}^{\mu_i - 1} \frac{(-1)^n}{n!\,(\lambda_i - \mu_i + 1) \cdots (\lambda_i - \mu_i + n)} \left(\frac{z}{2}\right)^{2n} = K_{\lambda_i - \mu_i}(z). \tag{37}$$

Using this relation, we substitute in (37) the functions $K_{\alpha_i \mu_i}(\alpha_j z)$ and their derivatives, in which either the λ_i or the μ_i are integers. We substitute for the remaining functions $K_{\lambda_i \mu_i}(\alpha_j z)$ and their derivatives using (35). Let r be the total degree of P in all the variables, and let q be the least common multiple of the denominators of all the numbers λ_i, μ_i $(i = 1, \ldots, l)$. Multiplying the relation obtained from (36) by $(\Pi_{i=1}^m \Pi_{j=1}^n v_{ij}) z^{2r}$ we transform it into a polynomial in $z^{1/q}$, $z^{-1/q}$, v_{ij}, v'_{ij}, w_{ij}, u_{ij}, s_{ij}, $K_{\tau_p}(\alpha_j z)$, $K'_{\tau_p}(\alpha_j z)$ $(i = 1, \ldots, l;\ p = 1, \ldots, m;\ j = 1, \ldots, n)$.

Let r_0 be the maximal degree of this polynomial in $z^{-1/q}$; multiplying by $z^{r_0/q}$, we eliminate the argument $z^{-1/q}$, and we let P_1 denote the resulting polynomial in $z^{1/q}$, v_{ij}, v'_{ij}, w_{ij}, u_{ij}, $K_{\tau_p}(\alpha_j z)$, $K'_{\tau_p}(\alpha_j z)$. Further, transforming P_1 using the method in Lemma 5 of [2], we obtain an algebraic relation between z, v_{ij}, v'_{ij}, s_{ij}, u_{ij}, w_{ij}, $K_{\tau_p}(\alpha_j z)$, $K'_{\tau_p}(\alpha_j z)$ $(i = 1,\ldots,l;\ p = 1,\ldots,m;\ j = 1,\ldots,n)$; but this contradicts Lemma 5, if in this lemma we set $\xi_j = \alpha_j$ and $\eta_i = \lambda_i - \mu_i$.

Bibliography

1. Carl Ludwig Siegel, *Über einige Anwendungen diophantischer Approximation*, Abh. Preuss. Akad. Wiss. Berlin Phys.-Math. Kl. **1929**, no. 1 (1930).

2. A. B. Sidlovskiĭ, *Transcendentality and algebraic independence of the values of entire functions of certain classes*, Moskov. Gos. Univ. Učen. Zap. No. 186 (1959), 11–70. (Russian)

3. ______, *Transcendence and algebraic independence of values of E-functions satisfying linear nonhomogeneous differential equations of second order*, Dokl. Akad. Nauk SSSR **169** (1966), 42–45; English transl. in Soviet Math. Dokl. **7** (1966).

4. ______, *On criterion for algebraic independence of values of a class of entire functions*, Izv. Akad. Nauk SSSR Ser. Mat. **23** (1959), 35–66; English transl. in Amer. Math. Soc. Transl. (2) **22** (1962).

Translated by N. KOBLITZ

Amer. Math. Soc. Transl.
(2) Vol. **119**, 1983

On the Estimation and Applications
of Character Sums*

WANG YUAN

§1. Introduction

Using Weil's contribution [1] concerning the Riemann hypothesis on finite algebraic function fields, Burgess [2] first introduced a method of estimating character sums: for the case of real primitive characters mod p (prime) he improved the well-known result of Pólya. Afterwards, he and the present author generalized and improved his results and obtained a number of applications. The final form of his method can be described as follows:

THEOREM A [6]. *Let χ be a primitive character* mod k, *and let η and r be respectively an arbitrarily given positive number and a positive integer. If k has no square factors or $r = 2$, then for any pair of integers N, H ($H > 0$)*

$$\left| \sum_{n=N+1}^{N+H} \chi(n) \right| \leqslant c_1(r, \eta) H^{1-1/(r+1)} k^{1/4r+\eta}.$$

From this we have the following two corollaries:

COROLLARY 1 [6]. *If $\chi(n) = \left(\frac{f}{n}\right)$ is a real primitive character* mod f, *then*

$$\left| \sum_{n=1}^{H} \left(\frac{f}{n}\right) \right| \leqslant c_2(r, \eta) H^{1-1/(r+1)} f^{1/4r+\eta},$$

where $\left(\frac{f}{n}\right)$ is the Kronecker symbol.

COROLLARY 2 [4]. *If χ is a nonprincipal character* mod p, *then for $H > p^{1/4+\eta}$*

$$\left| \sum_{n=1}^{H} \chi(n) \right| < c_3(\eta) \frac{H}{p^{\eta^2/6}}.$$

1980 *Mathematics Subject Classification.* Primary 10G20.
* Translation of Shuzue Jinzhan **7** (1964), 78–83; MR **37** #5162.

The purpose of this paper is to apply these estimates to the problem of the least solution of Pell's equation and the problem of the least nth nonresidue mod p. Furthermore, this paper also discusses the least nth nonresidue under the generalized Riemann hypothesis.

Let d be a positive integer which is not a perfect square, $d \equiv 0$ or $1 \pmod 4$. Let the lattice point (x_0, y_0) $(x_0, y_0 > 0)$ be the solution of the Pell's equation $x^2 - dy^2 = 4$ so that $x_0 + \sqrt{d}\, y_0$ is a minimum. Let

$$\varepsilon = \left(x_0 + \sqrt{d}\, y_0 \right)/2.$$

THEOREM 1. *For any $\delta > 0$ there exists $c_4(\delta)$ such that when $d > c_4(\delta)$*

$$\ln \varepsilon < (1/4 + \delta)\sqrt{d}\, \ln d.$$

This result improves that of Hua Loo-Keng [7].

Let $n \geqslant 2$ and $n \mid (p - 1)$. If the congruence

$$x^n \equiv c \pmod p, \qquad 1 \leqslant x \leqslant p$$

has no solution, we call c an nth nonresidue mod p; otherwise we call c an nth residue mod p. Denote the least positive nth nonresidue by $N(p, n)$.

THEOREM 2. *Let δ be an arbitrarily given positive number. Then, for sufficiently large p,*
(i) $N(p, n) \leqslant p^{1/4e^{(n-1)/n}+\delta}$ $(n \geqslant 2)$,
(ii) $N(p, n) \leqslant p^{1/12}$ $(n \geqslant 21)$,
(iii) $N(p, n) \leqslant p^{(\ln \ln n + 2)/4 \ln n}$ $(n > e^{33})$.

(i), (ii) and (iii) are improvements of the results of Vinogradov [8] and Buhštab [9].

§2. Proof of Theorem 1

Let

$$\sigma(a) = \sum_{n=1}^{a} \left(\frac{d}{n} \right), \qquad K(d) = \sum_{n=1}^{\infty} \left(\frac{d}{n} \right)\frac{1}{n}.$$

LEMMA 1. *Let $r \geqslant 4$ be an arbitrary integer and $\tau = 1/r$. Then for $a \geqslant d^{1/4+\tau}$*

$$\left| \sigma(a) \right| \leqslant c_5(\tau)ad^{-\tau^2/6}.$$

PROOF. Let $d = fm^2$, where f is the fundamental discriminant. Then

$$\sigma(a) = \sum_{\substack{n=1}}^{a} \left(\frac{d}{n} \right) = \sum_{\substack{n=1 \\ (n,m)=1}}^{a} \left(\frac{f}{n} \right) = \sum_{n=1}^{a} \left(\frac{f}{n} \right) \sum_{k \mid (n,m)} \mu(k)$$

$$= \sum_{k \mid m} \mu(k)\left(\frac{f}{k} \right) \sum_{n \leqslant a/k} \left(\frac{f}{n} \right).$$

Hence

$$|\sigma(a)| \le \sum_{k|m} \left| \sum_{n \le a/k} \left(\frac{f}{n} \right) \right|.$$

From Corollary 1 we see that

$$|\sigma(a)| \le \sum_{k|m} c_6(\tau) \left(\frac{a}{k} \right)^{1-1/(r+1)} f^{1/4r+1/4r(r+1)} \le c_5(\tau) a^{1-1/(r+1)} d^{1/4r+1/2r(r+1)}$$

$$\le c_5(\tau) a d^{-1/4(r+1)-1/r(r+1)+1/4r+1/2r(r+1)} = c_5(\tau) a d^{-1/4r(r+1)} < c_5(\tau) a d^{-\tau^2/6}.$$

The lemma is proved.

LEMMA 2. *For any* δ, $0 < \delta < \frac{1}{2}$, *there exists* $c_7(\delta)$ *such that when* $d > c_7(\delta)$

$$K(d) < (1/4 + \delta)\ln d.$$

PROOF. We have

$$K(d) = \sum_{n=1}^{\infty} \frac{\sigma(n) - \sigma(n-1)}{n} = \sum_{n=1}^{\infty} \frac{\sigma(n)}{n(n+1)} = \sum_1 + \sum_2 + \sum_3.$$

Let $\tau = 1/r \le \delta/2 < 2\tau$, where r is an integer. Then

$$|\Sigma_1| \le \left| \sum_{1 \le n < d^{1/4+\tau}} \frac{\sigma(n)}{n(n+1)} \right| \le \sum_{1 \le n < d^{1/4+\tau}} \frac{1}{n+1}$$

$$< \int_1^{d^{1/4+\tau}} \frac{dt}{t} + \frac{1}{d^{1/4+\tau}} \le \left(\frac{1}{4} + \frac{\delta}{2} \right) \ln d + \frac{1}{d^{1/4+\delta/4}}.$$

Thus from Lemma 1 we see that

$$|\Sigma_2| = \left| \sum_{d^{1/4+\tau} \le n < d} \frac{\sigma(n)}{n(n+1)} \right| \le c_8(\delta) d^{-\delta^2/96} \sum_{1 \le n < d} \frac{1}{n+1} < c_8(\delta) d^{-\delta^2/96} \ln d.$$

Finally, by the theorem of Pólya we see that

$$|\sigma(a)| \le \sum_{k|m} \sqrt{f} \ln f < \sqrt{d} \ln d.$$

Hence

$$|\Sigma_3| \le \sqrt{d} \ln d \cdot \sum_{n \ge d} \frac{1}{n(n+1)} = \frac{\ln d}{\sqrt{d}}.$$

Collecting the above, we obtain the lemma.

Theorem 1 is a corollary of Lemma 2 (see [10]).

§3. Proof of Theorem 2

LEMMA 3 [9]. *Let* $\Psi(x, y)$ *denote the number of positive integers no greater than* x *and containing at most* y *prime factors. Then*

$$\Psi(x, x^{1/\alpha}) = \rho(\alpha)x + O\left(\frac{x}{\sqrt{\ln x}} \right),$$

where the constants related to O depend only on α, and

$$\rho(\alpha) = \begin{cases} 1, & \text{if } 0 < \alpha \leqslant 1; \\ 1 - \displaystyle\int_1^\alpha \frac{dz_1}{z_1} + \int_2^\alpha \int_1^{z_1-1} \frac{dz_1 dz_2}{z_1 z_2} - \cdots \\ \quad + (-1)^{[\alpha]} \displaystyle\int_{[\alpha]}^\alpha \int_{[\alpha-1]}^{z_1-1} \cdots \int_1^{z_{[\alpha-1]}-1} \frac{dz_1 \cdots dz_{[\alpha]}}{z_1 \cdots z_{[\alpha]}}, & \text{if } \alpha > 1. \end{cases}$$

Thus we see that $\rho(\alpha)$ is a continuous monotonically decreasing function, and, when $\alpha \geqslant 6$,

$$\rho(\alpha) > e^{-\alpha(\ln a + \ln \ln \alpha + (6 \ln \ln \alpha)/\ln \alpha)}.$$

LEMMA 4. *Let R denote the number of nth residues* $\bmod p$ *in the interval* $1 \leqslant c \leqslant H$. *If* $n \mid (p - 1)$, $n \geqslant 2$ *and* $H > p^{1/4+\eta}$ $(\eta > 0)$, *then* $R = H/n + /S$, *where* $|S| < c_9(\eta) H p^{-\eta^2/6}$.

PROOF. Let $\chi(x) = e^{2\pi i \, \mathrm{ind}\, x/n}$. Then from Corollary 2 we see that

$$R = \sum_{x=1}^H \frac{1}{n} \sum_{a=1}^n e^{2\pi i a \, \mathrm{Ind}\, x/n} = \frac{H}{n} + \frac{1}{n} \sum_{a=1}^{n-1} \sum_{x=1}^H \chi(x)^a = \frac{H}{n} + S,$$

where $|S| < c_9(\eta) H p^{-\eta^2/6}$. The lemma is proved.

LEMMA 5. *If* $n \geqslant 2$, $n \mid (p - 1)$ *and* $\rho(\alpha) > 1/n + \delta$ $(\delta > 0)$, *then there exists* $c_{10}(\alpha, \delta)$ *such that* $N(p, n) \leqslant p^{1/4\alpha}$ *when* $p > c_{10}(\alpha, \delta)$.

PROOF. Take $H = [p^{1/4+\eta}] + 1$ $(\eta > 0)$. If $N(p, n) > p^{1/4\alpha}$, then the positive integers that are no greater than H and contain at most $p^{1/4\alpha}$ prime factors are all nth residues. Define R as in Lemma 4; then, when $\eta = \eta(\delta)$ is sufficiently small,

$$R \geqslant \Psi(H, p^{1/4\alpha}) \geqslant \rho((1 + 5\eta)\alpha)H + O(H/\sqrt{\ln p})$$

$$> (1/n + \delta/2)H + O(H/\sqrt{\ln p}),$$

so from Lemma 4 we see that

$$H/n + O(Hp^{-\eta^2/6}) \geqslant (1/n + \delta/2)H + O(H/\sqrt{\ln p}),$$

where the constants related to O depend only on α and δ. When p is sufficiently large, the above formula is impossible. Thus the proof is evident.

Now we prove Theorem 2 as follows:

(i) Take $\alpha = e^{(n-1)/n-\tau}$ $(1/2 > \tau > 0)$; then

$$\rho(\alpha) \geqslant 1 - \ln \alpha = 1/n + \tau.$$

Hence there exists $c_{11}(n, \tau)$ such that when $p > c_{11}(n, \tau)$ and $n \geqslant 2$

$$N(p, n) \leqslant p^{1/4e^{(n-1)/n}+\delta},$$

here $\delta > 0$ and $\lim_{\tau \to 0} \delta = 0$.

(ii) Take $\alpha = 3$. Then

$$\rho(3) = 1 - \ln 3 + \int_2^3 \int_1^{z_1-1} \frac{dz_1\,dz_2}{z_1 z_2} > 0.4804 > \frac{1}{21}.$$

Hence there exists c_{12} such that when $p > c_{12}$ and $n \geq 21$

$$N(p, n) \leq p^{1/12}.$$

(iii) Take $\alpha = (\ln n)/(\ln \ln n + 2)$. Then when $n > e^{33}$ we can always choose $\delta = \delta(n) > 0$ so that $\rho(\alpha) > 1/n + \delta$. Hence there exists $c_{13}(n)$ such that when $p > c_{13}(n)$ and $n > e^{33}$

$$N(p, n) \leq p^{(\ln \ln n + 2)/4 \ln n}.$$

§4. Conditional result

LEMMA 6 [4, 11]. *Under the generalized Riemann conjecture*,

$$\sum_{n=1}^{\infty} \Lambda(n)\chi(n)e^{-n/x} = \begin{cases} x + O(x^{1/2}\ln p), \\ \qquad \text{if } \chi \text{ is the principal character } \chi_0 \bmod p, \\ O(x^{1/2}\ln p), \text{ if } \chi \text{ is a nonprincipal character}; \end{cases}$$

here and afterwards the constants related to O *are absolute constants.*

THEOREM 3. *Under the generalized Riemann conjecture*

$$N(p, n) = O(\ln^2 p) \qquad (n \geq 2).$$

PROOF. It is well known that when $x \geq 2$, there is $c_{14} > 0$ such that

$$\Psi(x) = \sum_{n \leq x} \Lambda(n) \leq c_{14}x.$$

With summation by parts we see that when $c_{15} > 1$

$$\sum_{n > c_{15}x} \Lambda(n)e^{-n/x} \leq c_{14}e^{-c_{15}}(c_{15} + 1)x.$$

Consider the function

$$R(x) = \sum_{r_n \leq c_{15}x} \chi_0(r_n)\Lambda(r_n)e^{-r_n/x},$$

where $\sum_{r_n \leq c_{15}x}$ represents the sum of all positive nth nonresidues $\leq c_{15}x$. Then

$$R(x) \geq \sum_{r_n \geq 1} \chi_0(r_n)\Lambda(r_n)e^{-r_n/n} - \sum_{m > c_{15}x} \Lambda(m)e^{-m/x}$$

$$\geq \sum_{m=1}^{\infty} \chi_0(m)\Lambda(m)e^{-m/x}\left(1 - \frac{1}{n}\sum_{a=1}^{n} e^{2\pi i a\,\mathrm{Ind}\,m/n}\right) - c_{14}e^{-c_{15}}(c_{15} + 1)x$$

$$= \left(1 - \frac{1}{n}\right)\sum_{m=1}^{\infty} \chi_0(m)\Lambda(m)e^{-m/x} - \frac{1}{n}\sum_{a=1}^{n-1}\sum_{m=1}^{\infty} \Lambda(m)\chi_0(m)e^{2\pi i a\,\mathrm{Ind}\,m/n}e^{-m/x}$$

$$-c_{14}e^{-c_{15}}(c_{15} + 1)x \geq (1 - 1/n - c_{14}e^{-c_{15}}(c_{15} + 1))x + O(x^{1/2}\ln p).$$

Take $x = c_{16} \ln^2 p$. When c_{15} and c_{16} are sufficiently large we see that $R(x) > 0$. In other words,

$$N(p, n) \leqslant c_{15} c_{16} \ln^2 p.$$

The theorem is proved.

BIBLIOGRAPHY

1. André Weil, *Sur les courbes algébriques et les variétés qui s'en déduisent*, Actualités Sci. Indust., No. 1041, Hermann, Paris, 1948.

2. D. A. Burgess, *The distribution of quadratic residues and non-residues*, Mathematika **4** (1957), 106–112.

3. Georg Pólya, *Über die Verteilung der quadratischen Reste und Nichtreste*, Nachr. Königl. Ges. Wiss. Göttingen Math.-Phys. Kl. **1918**, 20–29.

4. Wang Yuan, *On the least primitive root of a prime*, Acta Math. Sinica **9** (1959), 432–441; English transl., Scientia Sinica **10** (1962); reprinted in Chinese Math. Acta **9** (1967), 704–717.

5. D. A. Burgess, *On character sums and primitive roots*, Proc. London Math. Soc. (3) **12** (1962), 179–192.

6. ______, *On character sums and L-series*, Proc. London Math. Soc. (3) **12** (1962), 193–206.

7. Loo-keng Hua, *On the least solution of Pell's equation*, Bull. Amer. Math. Soc. **48** (1942), 731–735.

8. I. M. Vinogradov, *On a bound for the least nonresidue of nth degree*, Izv. Akad. Nauk SSSR (6) **20** (1926), 47–58; reprinted in his *Selected works*, Izdat. Akad. Nauk SSSR, Moscow, 1952, pp. 81–88. (Russian)

9. A. A. Buhštab, *On those numbers in an arithmetic progression all prime factors of which are small in order of magnitude*, Dokl. Akad. Nauk SSSR **67** (1949), 5–8. (Russian)

10. Hua Loo-keng, "Binary quadratic forms (Siegel's theorem on the class number)", Chapter 12 in his *Introduction to number theory*, Science Publ. Co., Peking, 1957. (Chinese)

11. N. C. Ankeny, *The least quadratic non residue*, Ann. of Math. (2) **55** (1952), 65–72.

Translated by H. S. SUN

L-functions Of A Local Field, And A Real Quadratic Field*

I. Š. SLAVUTSKIĬ

In a series of papers [1–6] Leopoldt constructed a local analogue of formulas for the number of divisor classes of an absolutely abelian real field. In this regard the role of $L(1/\chi)$, the values of Dirichlet's L-functions when $s = 1$ for an even character χ, is filled by the values $\mathfrak{L}_p(\chi)$ containing the p-adic logarithm. Also, in the paper [6], written with Kubota, Leopoldt introduced local L-functions, investigated some of their properties and communicated that later he would prove the agreement of the L-functions at $s = 1$ and the values $\mathfrak{L}_p(\chi)$ for the corresponding characters, and also the absence of zeros of the local L-functions when $s = 1$ for nonprincipal even characters.

The author is unaware of further publications in this direction; moreover, it is possible to show that the introduction of a local L-function in the sense of Kubota and Leopoldt requires a refinement, without which the proof of the agreement of the L-function at $s = 1$ and $\mathfrak{L}_p(\chi)$ would be impossible. In this note we consider the case of a real quadratic field.

The Kubota-Leopoldt local L-function, defined as a certain bounded linear functional on a normed ring of convergent power series (on the group of principal units of a local field with $p \neq 2$), differs from the one considered in this note only by a factor of $(1 - \chi(p)p^{-1})^{-1}$. Namely, throughout this note

$$L(s\,|\,\chi) = (1 - \chi(p)/p)^{-1} L_{\mathrm{KL}}(s\,|\,\chi),$$

where $L_{\mathrm{KL}}(s\,|\,\chi)$ is the Kubota-Leopoldt L-function.

It is evident that $L(s\,|\,\chi) = L_{\mathrm{KL}}(s\,|\,\chi)$ if $p\,|\,f$, where f is the conductor modulus of χ. If $\chi = \varepsilon$, the principal character, with $f = 1$ and the Kubota-Leopoldt local zeta function has residue $1 - \frac{1}{p}$ at $s = 1$, then the function $L(s\,|\,\chi)$ introduced here has residue equal to unity.

1980 *Mathematics Subject Classification.* Primary 12A25, 12A50, 12A35, 12A70, 12B30.
*Translation of Izv. Vysš. Učebn. Zaved. Matematika **1969**, no. 2 (81), 99–105; MR **40** #4240.

1. Let $k = Q(\sqrt{d})$ be a real quadratic field with fundamental discriminant $d > 0$ and fundamental unit $E_1 = T_1 + U_1\sqrt{d}$, where T_1 and U_1 are integral or half-integral rational numbers, and Q is the field of rational numbers. The only nonprincipal character associated with the field k is the Kronecker symbol, and its conductor modulus is $f = d$. The generalized Bernoulli numbers $B^n(\chi)$ belonging to the character χ with conductor modulus f are defined by the identity

$$\frac{\sum_{r=1}^{f} \chi(r)e^{rt}t}{e^{ft} - 1} = \sum_{n=0}^{\infty} B^n(\chi)\frac{t^n}{n!}, \qquad |t| < \frac{2\pi}{f}.$$

Throughout the rest of this paper p stands for an odd prime.

We state as lemmas a number of results proved earlier by various authors.

LEMMA 1 ([7], p. 124; [12], §2). *If $p \nmid f$, $f > 1$ and $m \equiv n \pmod{(p-1)p^{l-1}}$, $l < \min\{m, n\}$, then the generalized Bernoulli numbers satisfy the Kummer congruence*

$$B^m(\chi)/m \equiv B^n(\chi)/n \pmod{p^l}. \tag{1}$$

LEMMA 2 ([8], §3). *Let $E_l = T_l + U_l\sqrt{d} = E_1^{p^{l-1}}$ when $p \mid d$ and $\overline{E}_l = \overline{T}_l + \overline{U}_l\sqrt{d} = E_1^{(p-\chi(p))p^{l-1}}$ when $p \nmid d$.*

Then the p-integral numbers U_l/p^{l-1} and U_1 are divisible by the same powers of the prime $p > 3$, and $\overline{U}_l/p^l$ and $\overline{U}_1/p$ by the same powers of any prime p.

LEMMA 3 ([9], p. 1191). *The number h of divisor classes of the real quadratic field k satisfies the congruences*

$$2h(U_l/T_l p^{l-1}) \equiv -(B^r(\chi)/r) \pmod{p^l}, \qquad d = np, n \geq 1, \chi \bmod n, \tag{2}$$

$$2h(\overline{U}_l/\overline{T}_l p^l) \equiv -(B^{2r}(\chi)/2r) \pmod{p^l}, \qquad p \nmid d, \chi \bmod d. \tag{3}$$

Here $r = (p-1)p^{l-1}/2$, and l is any natural number.

Suppose the character ω is defined as the p-adic limit

$$\omega = \omega(x) = \lim_{l \to \infty} x^{p^{l-1}}$$

for integers x with $(x, p) = 1$, so that $\omega(x) \in Q_p$, Q_p the field of p-adic numbers, $\omega^{p-1} = 1$ and $\omega(x) \equiv x \pmod{p}$. Then, taking into account the connection between the function defined here and the Kubota-Leopoldt L-function, we obtain

LEMMA 4 ([6], p. 336). *For a natural number m and nonprincipal character $\chi \neq \varepsilon$ the local L-function continuous on the ring of p-adic integers satisfies the equality*

$$L(1 - m \mid \chi) = -(B(\chi\omega^{-m})/m)(1 - \chi\omega^{-m}(p)p^{m-1})(1 - \chi(p)p^{-1})^{-1}. \tag{4}$$

It follows from (4) that

$$L(1 \mid \chi) = \lim_{|m| \to 0} L(1 - m \mid \chi) = \lim_{|m| \to 0} \{-(B^m(\chi\omega^{-m})/m)(1 - \chi(p)p^{-1}\}^{-1},$$

$$\tag{5}$$

where $|m|_p = |m|$ is the p-adic norm of m. The intrinsic meaning of the existence of the p-adic limit (5) consists in the presence of the Kummer congruence (1), from which Kubota and Leopoldt actually constructed the local L-function.

When $(x, p) = 1$ and $r = (p - 1)p^{l-1}/2$, then since $x^r \equiv \left(\frac{x}{p}\right)$ (mod p^l) and $x^{2r} \equiv 1$ (mod p^l), where $\chi_p(x) = \left(\frac{x}{p}\right)$ is the Legendre symbol, we have the p-adic limits $\lim_{l\to\infty} x^r = \left(\frac{x}{p}\right)$ and $\lim_{l\to\infty} x^{2r} = 1$. Hence from (5), in particular, we obtain

$$L(1|\chi) = \lim_{l\to\infty} \left\{ -\left(B^r\left(x\chi_p^{-1}\right)/r\right)\left(1 - \chi(p)p^{-1}\right)^{-1}\right\} \tag{6}$$

and

$$L(1|\chi) = \lim_{l\to\infty} \left\{ -\left(B^{2r}(\chi)/2r\right)\left(1 - \chi(p)p^{-1}\right)^{-1}\right\}. \tag{7}$$

2. The Dirichlet formula for the class number $h = h(d)$ of the real quadratic field $k = Q(\sqrt{d})$ has the form

$$E_1^{2h} = \prod_{\substack{(x,d)=1, \\ 0<x<d}} \left(1 - \zeta^x\right)^{-\chi(x)}, \qquad \zeta = \exp\frac{2\pi i}{d}.$$

Consider first the case $d = p$, where the divisor $\mathfrak{p} = (1 - \rho)$, $\rho = \exp(2\pi i/p)$, is a prime in the field $K = Q(\rho)$, and introduce the $\mathfrak{p}$-adic logarithm in the field $K_\mathfrak{p}$. For convenience we will henceforth assume that $\text{ord}_\mathfrak{p}\, p = 1$, and the logarithm will be considered for any local unit (not only principal!) by extension: $\log x = \frac{1}{k}\log x^k$ with $\text{ord}_\mathfrak{p}(x^k - 1) > 0$ and some natural number k (see [5]). In this case the local analogue of the Dirichlet formula has the form

$$2h\frac{\log E_1}{\sqrt{p}} = \frac{\tau(\chi)}{p}\frac{1}{p-1}\sum_{x \bmod p}\left(-\chi(x)\log\frac{(1-\rho^x)^{p-1}}{-p}\right),$$

where the summation extends over the complete system of least positive residues modulo p, $\chi(x)$ is the Legendre symbol, and the Gaussian sum $\tau(\chi)$ is equal to $\sqrt{p}$ or, briefly,

$$2h\left(\log E_1/\sqrt{p}\right) = \mathfrak{L}_p(\chi). \tag{8^1}$$

In the given case, in order to obtain (8^1), and likewise (8^2) and (8^3), it is not necessary to bring in some subgroup of the group of all units of the field (formalen Kreiseinheitenkern) as for an arbitrary absolutely abelian field (see [1] and [5]). Indeed,

$$\prod_{x \bmod p}\left(1 - \rho^x\right)^{-\chi(x)} = \left[\frac{\prod_x\left(1 - \rho^x\right)^{-\chi(x)(p-1)}}{(-p)^{-\Sigma_{x\bmod p}\chi(x)}}\right]^{1/(p-1)}$$

$$= \left[\prod_x\left\{\frac{(1-\rho^x)^{p-1}}{-p}\right\}^{-\chi(x)}\right]^{1/(p-1)}, \quad \text{where } \frac{(1-p^x)^{p-1}}{-p} \equiv 1 \,(\text{mod } \mathfrak{p}),$$

in view of the fact that

$$\frac{(1-\rho^x)^{p-1}}{-p} = -\prod_{1\leq\mu\leq p-1}\frac{1-\rho^x}{1-\rho^{x\mu}} = -\prod_{\substack{\mu' \\ \mu\mu'\equiv 1(\mathrm{mod}\ p)}} \left(1+\rho^{x\mu}+\cdots+\rho^{x\mu(\mu'-1)}\right)$$

$$\equiv -\prod_{\mu'}\mu' \equiv -(p-1)! \equiv 1\ (\mathrm{mod}\ \mathfrak{p}),$$

if we take into account that $\rho^x \equiv 1$ (mod $\mathfrak{p}$). Hence taking the logarithm of the Dirichlet formula leads to (8^1), if only we observe that $\mathrm{ord}_\mathfrak{p}\ E_1 = 0$.

Now suppose the discriminant of the field k has the form $d = np$, $n > 1$. We again designate a prime divisor $\mathfrak{p}\,|\,p$ in $K = Q(\zeta)$, $\zeta = \exp(2\pi i/d)$. If we put $\xi = \exp(2\pi i/n)$ and $\rho = \exp(2\pi i/p)$, then since $(x, d) = 1$ and $(n, p) = 1$ there exist integers a and b such that $an + bp = 1$ and

$$1 - \zeta^x = 1 - \xi^{xb}\rho^{xa} = 1 - \rho^{xa} + \rho^{xa}\left(1 - \xi^{xb}\right) \not\equiv 0\ (\mathrm{mod}\ \mathfrak{p}).$$

Hence in the same case where the discriminant of the field is composite we also obtain

$$2h\frac{\log E_1}{\sqrt{d}} = \frac{\tau(\chi)}{d}\sum_{x\,\mathrm{mod}\,d}\left(-\chi(x)\log(1-\zeta^x)\right),$$

where $\chi(x) = (\frac{d}{x})$ is the Kronecker symbol and $\tau(\chi) = \sqrt{d}$, or, briefly,

$$2h\left(\log E_1/\sqrt{d}\right) = \mathfrak{L}_p(\chi). \tag{8^2}$$

Finally, if $p\nmid d$, then $1 - \zeta \not\equiv 0$ (mod $\mathfrak{p}$), $\zeta = \exp(2\pi i/d)$, $\mathfrak{p}\,|\,p$ in $K = Q(\zeta)$, and $\mathrm{ord}_\mathfrak{p}\ E_1 = 0$; hence from Dirichlet's formula, putting, as above,

$$\mathfrak{L}_p(\chi) = \frac{\tau(\chi)}{d}\sum_{x\,\mathrm{mod}\,d}\left(-\chi(x)\log(1-\zeta^x)\right),$$

we conclude that

$$2h\left(\log E_1/\sqrt{d}\right) = \mathfrak{L}_p(\chi). \tag{8^3}$$

Thus, combining (8^1)–(8^3) and retaining the above values for $\mathfrak{L}_p(\chi)$, we finally obtain the local analogue of Dirichlet's formula for the number of divisor classes of the real quadratic field in the form

$$2h\left(\log E_1/\sqrt{d}\right) = \mathfrak{L}_p(\chi). \tag{8}$$

3. We will now show that for any real quadratic field (or equivalently, for any nonprincipal real even character)

$$L(1\,|\,\chi) = \mathfrak{L}_p(\chi), \tag{9}$$

where

$$L(1\,|\,\chi) \neq 0. \tag{10}$$

For the proof it is convenient to distinguish the same three cases as above.

1) Suppose $d = p$. Then in (2) we have $n = 1$ and $\chi = \varepsilon$, so that (2) assumes the form

$$2h\big(U_l/T_l p^{l-1}\big) \equiv -B^r(\varepsilon)/r \;(\mathrm{mod}\ p^l), \qquad (11)$$

where the $B^k(\varepsilon) = B_k$ are the usual Bernoulli numbers in the even numbering (in (11) the Bernoulli numbers are different from zero since $p \equiv 1\ (\mathrm{mod}\ 4)$). First of all, using an approximation to the logarithmic function in the form

$$\log x = \lim_{l \to \infty} \big((x^{p^l} - 1)/p^l\big)$$

for $\mathrm{ord}_{\mathfrak{p}}(x - 1) > 0$ ([10], p. 43; [11]), we obtain

$$\log E_1/\sqrt{p} = \lim_{l \to \infty} \big(U_l/T_l p^{l-1}\big). \qquad (12)$$

Indeed, we first observe that Lemma 2 implies $U_l \equiv 0\ (\mathrm{mod}\ p^{l-1})$; hence from $T_l^2 - U_l^2 p = -1$ it follows that $T_l \equiv 1\ (\mathrm{mod}\ p^{2l-1})$. Thus,

$$\frac{\log E_1}{\sqrt{p}} = \frac{1}{4}\frac{\log E_1^4}{\sqrt{p}} = \frac{1}{4\sqrt{p}} \lim_{l \to \infty} \frac{E_l^{4p^{l-1}} - 1}{p^{l-1}} = \frac{1}{4\sqrt{p}} \lim_{l \to \infty} \frac{E_l^4 - 1}{p^{l-1}}$$

$$= \frac{1}{4\sqrt{p}} \lim_{l \to \infty} \left[\frac{T_l^4 - 1}{p^{l-1}} + 4\frac{U_l T_l^3}{p^{l-1}}\sqrt{p} + 6\frac{U_l^2 T_l^2}{p^{l-1}}p + \cdots \right] = \lim_{l \to \infty} \frac{U_l}{T_l p^{l-1}}.$$

Since in this case $\chi = \chi_p$ is the Legendre symbol, from (6), inasmuch as $\chi(p) = 0$, we obtain

$$L(1\,|\,\chi) = \lim_{l \to \infty} \{-B^r(\varepsilon)/r\},$$

or, along with (11) and (12), finally

$$2h\big(\log E_1/\sqrt{p}\,\big) = L(1\,|\,\chi). \qquad (13^1)$$

From (13^1) and (8) we obtain (9). Since for $p > 3$ we have, by Lemma 2,

$$\mathrm{ord}_{\mathfrak{p}}\big(U_l/T_l p^{l-1}\big) = \mathrm{ord}_{\mathfrak{p}}\, U_1$$

and, as is evident from (13^1), the nonvanishing of $L(1\,|\,\chi)$ is equivalent to the nonvanishing of the local regulator $\log E_1$ of the field $Q(\sqrt{d}\,)$, we conclude that (12) implies (10).

2) In the case $d = np$ the equality

$$\lim_{l \to \infty} \big(U_l/T_l p^{l-1}\big) = \log E_1/\sqrt{d}$$

is obtained in exactly the same way as (12). In (6), $\chi\chi_p^{-1}(x) = \left(\frac{\eta^n}{x}\right)$, where $\eta = (-1)^{(p-1)/2}$, inasmuch as $\left(\frac{d}{x}\right) = \left(\frac{\eta^n}{x}\right)\left(\frac{np}{x}\right)$, so that passage to the limit in (2) again gives

$$2h\big(\log E_1/\sqrt{d}\,\big) = L(1\,|\,\chi), \qquad (13^2)$$

whence, as above, $\mathfrak{L}_p(\chi) = L(1\,|\,\chi)$ for all p, and $L(1\,|\,\chi) \neq 0$ for $p > 3$.

3) Finally, if $p \nmid d$, then, taking into account Lemma 2 and the fact that $\overline{T}_l^2 - \overline{U}_l^2 d = +1$, we first conclude that $\overline{T}_l^2 \equiv 1 \pmod{p^{2l}}$. Hence

$$
\frac{\log E_1}{\sqrt{d}} = \frac{d}{2(p - \chi(p))\sqrt{d}} \log \overline{E}_1^2 = \frac{1}{2(p - \chi(p))\sqrt{d}} \lim_{l \to \infty} \frac{\overline{E}_1^{2p^{l-1}} - 1}{p^{l-1}}
$$

$$
= \frac{1}{2(p - \chi(p))\sqrt{d}} \lim_{l \to \infty} \frac{\overline{E}_l^2 - 1}{p^{l-1}} = \frac{p}{2(p - \chi(p))\sqrt{d}} \lim_{l \to \infty} \frac{\overline{E}_l^2 - 1}{p^l}
$$

$$
= \frac{1}{2(1 - \chi(p)p^{-1})\sqrt{d}} \lim_{l \to \infty} \left[\frac{\overline{T}_l^2 - 1}{p^l} + 2\overline{T}_l \frac{\overline{U}_l}{p^l} \sqrt{d} + \frac{\overline{U}_l^2 d}{p^l} \right]
$$

$$
= \frac{1}{(1 - \chi(p)p^{-1})} \lim_{l \to \infty} \frac{\overline{U}_l}{\overline{T}_l p^l},
$$

or, finally,

$$
\log E_1/\sqrt{d} = \lim_{l \to \infty} \overline{U}_l/\overline{T}_l p^l (1 - \chi(p)p^{-1})^{-1}. \tag{14}
$$

Then, multiplying (3) by $(1 - \chi(p)p^{-1})^{-1}$ and passing to the limit as $l \to \infty$, we obtain with the aid of (7) and (14)

$$
2h\left(\log E_1/\sqrt{d}\right) = L(1 \mid \chi), \tag{13^3}
$$

so that along with (8^3), since

$$
\operatorname{ord}_\mathfrak{p}\left(\overline{U}_1/p\right) = \operatorname{ord}_\mathfrak{p}\left(\overline{U}_l/p^l\right),
$$

we have in this case also (9) and (10).

4. We remark that there is another way of proving (9). The values $\mathfrak{L}_p(\chi)$ and $L(1 \mid \chi)$ are approximated, respectively, by expressions $B^n(\mathfrak{L}_p)$ and $B^n(L)$ depending on the Bernoulli numbers $B^n(\chi)$; it turns out that for any $\delta > 0$ we have in Q_p for a sufficiently large natural number n the inequality $|B^n(\mathfrak{L}_p) - B^n(L)| < \delta$. Then for sufficiently large n we obtain

$$
|\mathfrak{L}_p(\chi) - L(1 \mid \chi)| \leqslant |\mathfrak{L}_p(\chi) - B^n(\mathfrak{L}_p)| + |B^n(\mathfrak{L}_p) - B^n(L)|
$$
$$
+ |B^n(L) - L(1 \mid \chi)| < 3\delta,
$$

and hence $\mathfrak{L}_p(\chi) = L(1 \mid \chi)$. However, this approach (which probably carries over to any character χ with $\chi(-1) = +1$) is tedious and does not touch the question of the nonvanishing of $L(1 \mid \chi)$.[1]

Leningrad   Received 6/FEB/68

---

[1] *Added in proof.* After this paper was written there appeared a paper of Brumer [13], in which he proved, using results of Baker, the nonvanishing of the p-adic regulator of an absolutely abelian real field, and hence the values $\mathfrak{L}_p(\chi)$. Using this fact, Fresnel [14] obtained by the methods of p-adic interpolation that $L(1 \mid \chi) = \mathfrak{L}_p(\chi)$, exactly as indicated in the last section of our paper.

BIBLIOGRAPHY

1. Heinrich-Wolfgang Leopoldt, *Über Einheitengruppe und Klassenzahl reeller abelscher Zahlkörper*, Abh. Deutsch. Akad. Wiss. Berlin Kl. Math. Nat. **1953**, no. 2 (1954).

2. ______, *Eine Verallgemeinerung der Bernoullischen Zahlen*, Abh. Math. Sem. Univ. Hamburg **22** (1958), 131–140.

3. ______, *Über Klassenzahlprimteiler reeller abelscher Zahlkörper als Primteiler verallgemeinerter Bernoullischer Zahlen*, Abh. Math. Sem. Univ. Hamburg **23** (1959), 36–47.

4. ______, *Über Fermatquotienten von Kreisenheiten und Klassenzahlformeln modulo p*, Rend. Circ. Mat. Palermo (2) **9** (1960), 39–50.

5. ______, *Zur Arithmetik in abelschen Zahlkörpern*, J. Reine Angew. Math. **209** (1962), 54–71.

6. Tomio Kubota and Heinrich-Wolfgang Leopoldt, *Eine p-adische Theorei der Zetawerte*. I, J. Reine Angew. Math. **214/215** (1964), 328–339.

7. I. Š. Slavutskiĭ, *Generalized Voronoĭ congruence and the number of classes of ideals of an imaginary quadratic field*. II, Izv. Vysš. Učebn. Zaved. Matematika **1966**, no. 4(53), 118–126. (Russian)

8. ______, *Upper bounds and numerical calculation of the number of ideal classes of real quadratic fields*, Izv. Vysš. Učebn. Zaved. Matematika **1965**, no. 2(45), 161–165; English transl. in Amer. Math. Soc. Transl. (2) **82** (1969).

9. A. A. Kiselev and I. Š. Slavutskiĭ, *On the number of classes of ideals of a quadratic field and its rings*, Dokl. Akad. Nauk SSSR **126** (1959), 1191–1194. (Russian)

10. I. Š. Slavutskiĭ, *Arithmetical properties of the number of classes of ideals of quadratic fields*, Candidate's Dissertation, Leningrad, 1959. (Russian)

11. Heinrich-Wolfgang Leopoldt, *Zur Approximation des p-adischen Logarithmus*, Abh. Math. Sem. Univ. Hamburg **25** (1961), 77–81.

12. A. A. Kiselev and I. Š. Slavutskiĭ, *Certain congruences for the number of representatives by sums of an odd number of squares*, Dokl. Akad. Nauk SSSR **143** (1962), 272–274; English transl. in Soviet Math. Dokl. **3** (1962).

13. Armand Brumer, *On the units of algebraic number fields*, Mathematika **14** (1967), 121–124.

14. Jean Fresnel, *Nombres de Bernoulli et fonctions L p-adiques*, Ann. Inst. Fourier (Grenoble) **17** (1967), fasc. 2, 281–333.

15.* Kenkichi Iwasawa, *Lectures on p-adic L-functions*, Ann. of Math. Studies, No. 74, Princeton Univ. Press, Princeton, N.J., 1972.

Translated by G. A. KANDALL

---

*Added by translator.

Amer. Math. Soc. Transl.
(2) Vol. **119**, 1983

Bilinear Forms and Categories of Representations*

S. A. OVSIENKO AND A. V. ROĬTER

A quiver Γ consists of a set of points Γ_0 and a set of arrows Γ_1, each of which joins two (not necessarily distinct) points. We say that a representation of Γ over a field K is defined if there is associated to each $\gamma \in \Gamma_0$ a finite-dimensional space V_γ over K and to each edge $l \in \Gamma_1$ from γ_1 to γ_2 ($\gamma_1, \gamma_2 \in \Gamma_0$) a K-linear operator $f_l\colon V_{\gamma_1} \to V_{\gamma_2}$. By the dimension of the representation V we mean the vector $\dim V = (x_\gamma)$, $\gamma \in \Gamma_0$, where $x_\gamma = \dim_K V_\gamma$.

For each quiver Γ there is a certain associative algebra $\Lambda(\Gamma)$ over K. A basis of $\Lambda(\Gamma)$ over K consists of the directed paths in Γ together with idempotents corresponding to the points $\gamma \in \Gamma_0$. Multiplication in $\Lambda(\Gamma)$ is defined in the natural way. If Γ_0 and Γ_1 are finite and Γ contains no loops or directed cycles, then $\Lambda(\Gamma)$ is finite dimensional. The category of representations of a quiver Γ (the concepts of morphism, direct sum, etc. for quiver representations are defined in the natural way [1]) is equivalent in this case to the category of finitely generated right $\Lambda(\Gamma)$-modules. It is easy to show that $\Lambda(\Gamma)$ is hereditary [9].

The quivers having, to within isomorphism, a finite number of indecomposable representations (quivers of finite type) were described in [1] and [3]. Gabriel [2] showed that quivers of finite type are the same as Dynkin diagrams without multiple bonds (the arrows in the diagram can be directed arbitrarily), the dimensions of their indecomposable representations coincide with the positive roots of the diagram, and in each dimension that is a root there is exactly one indecomposable representation. A direct proof of the coincidence of the dimensions of the indecomposable representations and the positive roots is given in [1], where, in this connection, for quiver representations there are defined reflection functors and Coxeter functors, which act on the dimensions of representations like reflections and Coxeter transformations in the sense of root systems. An essential role in the proof is played by the quadratic form of the quiver, $F(x) = \Sigma_{\gamma \in \Gamma_0} x_\gamma^2 - \Sigma_{l \in \Gamma_1} x_{\gamma_1} x_{\gamma_2}$, where γ_1 and γ_2 are the initial and terminal

---

1980 *Mathematics Subject Classification.* Primary 16A64, 15A63, 16A90.

* Translation of Matrix Problems, Inst. Mat. Akad. Nauk Ukrain. SSR, Kiev, 1977, pp. 71–80; MR **58** #28130.

points of the edge l; in particular, it follows from [2] and [6] that the dimensions of the indecomposable representations coincide with the nonnegative integral roots of the equation $F(x) = 1$. This form, which was introduced in [2], will be called the Tits form in the sequel. An analogous quadratic form arises in other matrix problems (see [4] and [8]).

Apparently, along with the above-mentioned quadratic form a significant role in the theory of matrix problems must also be played by a nonsymmetric bilinear form (whose quadratic form is the Tits form). Suppose U is an arbitrary differential graded category (DGC) [10] and W is a vector space whose basis vectors e_i are in one-to-one correspondence with the points of U. We denote by $h_{\alpha\beta}$ the number of unbroken arrows and by $s_{\alpha\beta}$ the number of broken arrows from α to β (in U; see [10]). Then we define a bilinear form $\langle \ , \ \rangle$ by means of the equalities $\langle e_\alpha, e_\beta \rangle = s_{\alpha\beta} - h_{\alpha\beta}$.[1] In particular, if for a quiver Γ we denote by $h_{\alpha\beta}$ the number of arrows going from α to β, then $\langle e_\alpha, e_\beta \rangle = -h_{\alpha\beta}$ $(\alpha \neq \beta)$, and $\langle e_\alpha, e_\beta \rangle = 1 - h_{\alpha\alpha}$. We will show that while the quadratic form determines the dimensions of the indecomposable objects of the category of representations of a quiver of finite type, the bilinear form determines the dimensions of sets of morphisms of this category. More precisely, we have

THEOREM 1. *If X and Y are indecomposable representations of a quiver Γ of finite type and $\langle \dim X, \dim Y \rangle \geqslant 0$, then*

$$\langle \dim X, \dim Y \rangle = |\operatorname{Hom}_\Lambda(X, Y)|$$

and $\operatorname{Ext}^1_\Lambda(X, Y) = 0$, while if $\langle \dim X, \dim Y \rangle < 0$, then

$$\langle \dim X, \dim Y \rangle = -|\operatorname{Ext}^1_\Lambda(X, Y)|$$

and $\operatorname{Hom}_\Lambda(X, Y) = 0$, where $\Lambda = \Lambda(\Gamma)$.[2]

To prove Theorem 1 we need the following result, which is of independent interest.

PROPOSITION 1. *Suppose Λ is an algebra of finite cohomological dimension and R is the category of its finitely generated right modules. Then the correspondence*

$$\langle X, Y \rangle \to \sum_{i=0}^{\infty} (-1)^i |\operatorname{Ext}^i_\Lambda(X, Y)| \qquad \left(\operatorname{Ext}^0_\Lambda = \operatorname{Hom}_\Lambda\right)$$

defines a bilinear form on $K(R) \otimes \mathbf{Q}$, where $K(R)$ is the Grothendieck group of the category R.

Indeed, suppose $0 \to A \to B \to C \to 0$ is an exact sequence in R. If we write down the exact homology sequences

$$0 \to \operatorname{Hom}_\Lambda(X, A) \to \operatorname{Hom}_\Lambda(X, B) \to \operatorname{Hom}_\Lambda(X, C) \to \operatorname{Ext}^1_\Lambda(X, A) \to \ldots,$$

$$0 \to \operatorname{Hom}_\Lambda(C, X) \to \operatorname{Hom}_\Lambda(B, X) \to \operatorname{Hom}_\Lambda(A, X) \to \operatorname{Ext}^1_\Lambda(C, X) \to \ldots$$

---

[1]After this paper was submitted for publication, the authors learned that Ringel [7] introduced an analogous form and proved an assertion equivalent to our Proposition 1.

[2]If W is a finite-dimensional vector space over K, then $|W|$ denotes the K-dimension of W.

and equate the alternating sums of the dimensions of their terms to zero, we obtain $\langle X, B \rangle = \langle X, A \rangle + \langle X, C \rangle$ and $\langle B, X \rangle = \langle A, X \rangle + \langle C, X \rangle$ for any $X \in R$, which implies Proposition 1.

We apply Proposition 1 to the algebra $\Lambda = \Lambda(\Gamma)$. The canonical mapping $R \to K(R)$ amounts to taking the dimension of the representation. It is now easy to verify that the bilinear form in Proposition 1 for the representations of a quiver coincides with the bilinear form for the corresponding DGC. In this case, the quadratic form constructed from the bilinear one coincides with the Tits form. In the case of quivers with relations (see [4]), the quadratic form constructed from the bilinear one in Proposition 1 coincides with the form of Brenner [4].

To prove Theorem 1 we need to construct a Coxeter functor (see [1] and [5]). We use the method of construction of C^+ given in [5]. If M is a right Λ-module, then $\mathrm{Ext}^1_\Lambda(M, \Lambda)$ is a left Λ-module. The functor

$$M \to \mathrm{Ext}^1_\Lambda(M, \Lambda)^* = \mathrm{Hom}_K\big(\mathrm{Ext}^1_\Lambda(M, \Lambda), K\big)$$

is called the Coxeter functor C^+, $C^+(M) = \mathrm{Ext}^1_\Lambda(M, \Lambda)^*$. Note that $\mathrm{Ext}^i_\Lambda(X, Y) = 0$, $i \geq 2$, and

$$\langle X, Y \rangle = |\mathrm{Hom}_\Lambda(X, Y)| - |\mathrm{Ext}^1_\Lambda(X, Y)|$$

for $\Lambda = \Lambda(\Gamma)$.

The functor C^+ acts on the full subcategory $\bar{R} \subset R$ of modules containing no projective direct summands like an isomorphism of $\bar{R}$ onto $C^+(\bar{R})$, i.e. if $X, Y \in \bar{R}$, then $\mathrm{Hom}_\Lambda(X, Y) = \mathrm{Hom}_\Lambda(C^+(X), C^+(Y))$. We will show that if $X, Y \in \bar{R}$, then $\mathrm{Ext}^1_\Lambda(X, Y) = \mathrm{Ext}^1_\Lambda(C^+(X), C^+(Y))$. It suffices to show that the term Z in an exact sequence $0 \to Y \overset{\sigma}{\to} Z \overset{\pi}{\to} X \to 0$ contains no projective direct summands if $X, Y \in \bar{R}$. But if $Z = P \oplus M$ with P projective and if $p: Z \to P$ is a projection, then $p\sigma = 0$ (since $\mathrm{Im}\, p\sigma \subset P$ is projective); hence P is a direct summand of X. Thus, $\langle X, Y \rangle = \langle C^+(X), C^+(Y) \rangle$. The functor C^+ induces a transformation C of the vector space $K(R) \otimes \mathbf{Q}$ such that $\dim(C^+(X)) = C(\dim X)$ if $X \in \bar{R}$. According to [1], for any $X \in R$ there exists $k \geq 0$ such that $C^k(\dim X)$ has negative coordinates. Consequently, for any indecomposable X there exists $k \geq 0$ such that $C^{+k}(X)$ is indecomposable and projective. So suppose $C^{+k}(X)$ and $C^{+l}(Y)$ are indecomposable and projective. If $k \leq l$, then

$$\langle X, Y \rangle = |\mathrm{Hom}_\Lambda(C^{+k}(X), C^{+k}(Y))| - |\mathrm{Ext}^1_\Lambda(C^{+k}(X), C^{+k}(Y))|$$

$$= |\mathrm{Hom}_\Lambda(X, Y)|.$$

If $k > l$, then

$$\langle X, Y \rangle = |\mathrm{Hom}_\Lambda(C^{+l}(X), C^{+l}(Y))| - |\mathrm{Ext}^1_\Lambda(C^{+l}(X), C^{+l}(Y))|.$$

For any nonzero homomorphism $\varphi: C^{+l}(X) \to C^{+l}(Y)$, $\mathrm{Im}\,\varphi$ is projective because Λ is hereditary. But $C^{+l}(X)$ is indecomposable; hence $\mathrm{Im}\,\varphi = C^{+l}(X)$, i.e. $l = k$, which contradicts the assumption $k > l$. Hence,

$$\langle X, Y \rangle = -|\mathrm{Ext}^1_\Lambda(C^{+l}(X), C^{+l}(Y))| = -|\mathrm{Ext}^1_\Lambda(X, Y)|.$$

The theorem is proved.

COROLLARY 1. *If X and Y are indecomposable representations of a quiver Γ of finite type, $X \neq Y$, and $\operatorname{Hom}_\Lambda(X, Y) \neq 0$, then $\operatorname{Hom}_\Lambda(Y, X) = 0$.*

Indeed, if $\operatorname{Hom}_\Lambda(Y, X) \neq 0$, then, by Proposition 1,

$$\langle \dim X - \dim Y, \dim X - \dim Y \rangle = \langle X, X \rangle + \langle Y, Y \rangle - \langle X, Y \rangle - \langle Y, X \rangle$$
$$= 2 - |\operatorname{Hom}_\Lambda(X, Y)| - |\operatorname{Hom}_\Lambda(Y, X)| \leqslant 0,$$

which contradicts the fact that the Tits form is positive definite.

If Γ is a quiver of finite type, then in each dimension there are, to within isomorphism, only finitely many representations. If we fix bases in the spaces V_γ, then each representation having dimension x can be defined by a set of matrices $M_{\alpha\beta}$ of size $x_\alpha \times x_\beta$, $\alpha, \beta \in \Gamma_0$, where α and β are the pairs of points for which there exists an arrow from α to β. These matrices form a vector space M of dimension $\Sigma_{\alpha,\beta} x_\alpha x_\beta$ (the sum extends over the pairs α, β described in the previous sentence). Two representations V and V' are isomorphic if and only if there exists a set of nonsingular matrices M_α ($\alpha \in \Gamma_0$) such that $M'_{\alpha\beta} = M_\alpha^{-1} M_{\alpha\beta} M_\beta$. Since Γ is of finite type, among the representations having dimension x there exists a representation V_0 such that the representations isomorphic to V_0 form an open set in M in the sense of Zariski (V_0 is a representation of general position). In the case where x is a root of the corresponding Dynkin diagram, a representation of general position is indecomposable; but otherwise V_0 is decomposable. If we decompose V_0 into indecomposable representations, we obtain a decomposition of the vector x into a sum of roots. It turns out that such a decomposition can be obtained from the bilinear form associated with the quiver Γ.

Consider a vector space U over the field $\mathbf{Q}$, and on this space consider a bilinear form $\langle \ , \ \rangle$ such that its associated quadratic form is integral (a quadratic form $(a_{ij})_{1 \leqslant i,j \leqslant n}$ is called integral if the numbers $2a_{ij}/a_{ii}$ are integral; such, in particular, is the bilinear form associated with a quiver of finite type). Let $U^+ \subset U$ denote the set of vectors with nonnegative coordinates. An integral vector $x \in U^+$ is called indecomposable if for any decomposition into a sum of two integral vectors, $x = a + b$, $a, b \in U^+$, either $\langle a, b \rangle < 0$ or $\langle b, a \rangle < 0$. A decomposition $x = \Sigma_1^t x_i$ is called canonical if $\langle x_i, x_j \rangle \geqslant 0$ and each x_i is indecomposable.

Note that a vector need not have a canonical decomposition for a given bilinear form; for example, if $f_{12} = f_{23} = -1$, $f_{11} = f_{22} = f_{33} = f_{31} = 1$ and $f_{21} = f_{32} = f_{13} = 0$, then the vector $x = (1, 1, 1)$ has no canonical decomposition.

THEOREM 2. *Suppose Γ is a quiver of finite type and $\langle \ , \ \rangle$ is the corresponding bilinear form. Then for any integral $x \in U^+$ there exists a unique canonical decomposition $x = \Sigma_1^t x_i$, and the set of indecomposable vectors is the same as the set of positive roots of the corresponding diagram.*

The proof contains a constructive method for finding such a decomposition.

Before beginning the proof we examine the behavior of bilinear forms under reflections. Recall that for a nonisotropic vector $e \in U$ there is defined a reflection transformation

$$w_e(x) = x - 2\frac{\langle x, e \rangle}{\langle e, e \rangle} e$$

with respect to this vector; if in some basis the form has the matrix (a_{ij}), $1 \leqslant i, j \leqslant n$, then the reflection with respect to a basis vector e_i is defined by

$$w_i(e_j) = e_j - 2\frac{a_{ij}}{a_{ii}} e_i.$$

Suppose we have a bilinear form with matrix (f_{ij}), $1 \leqslant i, j \leqslant n$, where the f_{ij} are integers, and the corresponding quadratic form is integral. Assume we have performed the reflection with respect to one of the basis vectors e_i. Let us calculate the matrix of the form in the new basis $\{w_i(e_j)\}$, $1 \leqslant j \leqslant n$:

$$f'_{jk} = \langle w_i(e_j), w_i(e_k) \rangle = \left\langle e_j - \frac{f_{ij} + f_{ji}}{f_{ii}} e_i, e_k - \frac{f_{ik} + f_{ki}}{f_{ii}} e_i \right\rangle$$

$$= f_{jk} + \frac{f_{ki} f_{ij} - f_{ji} f_{ik}}{f_{ii}},$$

in particular, $f'_{ik} = f_{ki}$ and $f'_{ki} = f_{ik}$. Thus, although the reflection preserves the quadratic form ($f_{jk} + f_{kj} = f'_{jk} + f'_{kj}$), it does not, in general, preserve the matrix of the bilinear form. Let us consider in more detail the case where $\langle \ , \ \rangle$ is the bilinear form associated with a quiver Γ. If a vertex i is such that there is no edge entering i (leaving i), then this vertex is called $(-)$-admissible $((+)$-admissible). If we perform a reflection in the vertex i, where i is $(+)$-admissible $((-)$-admissible), i.e. $f_{ij} = 0$ ($f_{ji} = 0$), $j = 1, \dots, n$, then in the new basis we have $f'_{ij} = f_{ji}$, $f'_{ji} = 0$ ($f'_{ji} = f_{ij}$, $f'_{ij} = 0$), and all other f_{jk} are unchanged. Thus, the resulting form corresponds to the quiver $F_i^+(\Gamma)$ ($F_i^-(\Gamma)$) obtained from Γ by reversing the directions of the edges entering i (leaving i). If the vertex i is neither $(+)$-admissible nor $(-)$-admissible, then the form obtained after reflection does not correspond to any quiver. For example, the form corresponding to the quiver $\overset{1}{\circ} \ \rightarrow \ \overset{2}{\circ}$ $\rightarrow \overset{3}{\circ}$ after reflection in the vertex 2 has coefficients $f'_{11} = f'_{22} = f'_{33} = f'_{31} = 1$, $f'_{21} = f'_{32} = f'_{13} = -1$ and $f'_{12} = f'_{23} = 0$.

This calculation explains the fact that under reflections in $(+)$-admissible or $(-)$-admissible vertices (the reflections are understood to be functors on the representations of the quiver; see the definition in [1]) the directions of the arrows entering i (or leaving i) in $F_i^{\pm}(\Gamma)$ are reversed, and the fact that reflection functors exist only for $(+)$-admissible or $(-)$-admissible vertices.

The above observations easily imply the following

LEMMA. *Suppose Γ is a quiver without loops, $\langle \ , \ \rangle_\Gamma$ is the associated form, α is a $(+)$-admissible vertex, and $x, y \in U$. Then*

$$\langle x, y \rangle_\Gamma = \langle w_\alpha(x), w_\alpha(y) \rangle_{F_\alpha^+(\Gamma)}.$$

Suppose Γ is a quiver of finite type. According to [1], there exists an ordering of the vertices $\alpha_1, \ldots, \alpha_n$ of Γ_0 such that α_1 is $(+)$-admissible in Γ, α_2 is $(+)$-admissible in $F_{\alpha_1}^+(\Gamma), \ldots, \alpha_n$ is $(+)$-admissible in $F_{\alpha_{n-1}}^+ \cdots F_{\alpha_2}^+ F_{\alpha_1}^+(\Gamma)$. The transformations $F = w_{\alpha_n} \cdots w_{\alpha_1}$ and $F^- = (F)^{-1}$ are called Coxeter transformations. Consider $x \in U$. There exists $k \geqslant 0$ such that $F^k(x)$ has nonpositive coordinates (see [1]).

We prove Theorem 2 by induction on n, the number of points in Γ_0. For $n = 1$, Γ is simply a point without arrows, and the existence and uniqueness of the decomposition are obvious. Suppose existence and uniqueness have been proved for any quiver of finite type containing fewer than n points, and suppose $|\Gamma_0| = n$. Note that if an integral vector $x \in U$ has a zero coordinate (we say that x is unfaithful at this coordinate), then, viewing x in a smaller subdiagram, we obtain by means of the induction assumption a canonical decomposition in the subdiagram. This decomposition, as is easily seen, is a canonical decomposition of x in the whole diagram of Γ, and Theorem 2 is obviously true in this case.

Consider the sequence Π: $x, w_{\alpha_1}(x), w_{\alpha_2}w_{\alpha_1}(x), \ldots, F(x), w_{\alpha_1}F(x), \ldots$; let $y = w_{\alpha_k} \cdots w_{\alpha_1}F^l(x) = w(x)$ be the first element of this sequence having a nonpositive coordinate (which is obviously α_k, since the reflection in the point α_k changes only the α_k coordinates of a vector z). Denote the value of this coordinate by t. Consider the vector $y - t\alpha_k$. This vector is unfaithful; hence it has a canonical decomposition $y - t\alpha_k = \Sigma_1^s y_j$ (relative to the form associated with $F_{\alpha_k}^+ \cdots F_{\alpha_1}^+(\Gamma)$). Thus,

$$y = t\alpha_k + \sum_{j=1}^s y_j, \qquad \langle y_i, y_j \rangle_{F_{\alpha_k}^+ \cdots F_{\alpha_1}^+(\Gamma)} \geqslant 0$$

(because the decomposition is canonical), and

$$\langle t\alpha_k, y_i \rangle_{F_{\alpha_k}^+ \cdots F_{\alpha_1}^+(\Gamma)} \geqslant 0, \qquad \langle y_i, t\alpha_k \rangle_{F_{\alpha_k}^+ \cdots F_{\alpha_1}^+(\Gamma)} \geqslant 0,$$

since $t \leqslant 0$, the vector y_i is unfaithful at the vertex α_k, and α_k is $(-)$-admissible in $F_{\alpha_k}^+ \cdots F_{\alpha_1}^+(\Gamma)$. Then $x = w^{-1}(y) = tw^{-1}(\alpha_k) + \Sigma_{j=1}^s w^{-1}(y_j)$ will be a canonical decomposition of x once we show that no summand of the right-hand side has negative coordinates (the inner products of pairs of summands will be nonnegative by the lemma, and the $w^{-1}(y_j)$ will be roots, since the y_j are roots by the induction assumption). Let $y_0 = t\alpha_k$ and consider the vectors

$$w_{\alpha_k}(y_i), \quad w_{\alpha_{k-1}}w_{\alpha_k}(y_i), \ldots, F^{-l}w_{\alpha_1} \cdots w_{\alpha_k}(y_i), \qquad i = 0, 1, \ldots, s.$$

Assume that $w'(y_j)$ $(w' = w_{\alpha_p} \cdots w_{\alpha_{k-1}}w_{\alpha_k}, \ \Gamma' = F_{\alpha_{p-1}}^+ \cdots F_{\alpha_1}^+(\Gamma))$ is the first among them with a negative coordinate. Since y_j is a root, we have $w'(y_j) = -\beta$, where $\beta = \alpha_p$ is a simple root (since in a root there can be coordinates of only one sign). But according to the lemma,

$$\left\langle w'(y_j), w'(y_i) \right\rangle_{\Gamma'} = \left\langle -\beta, w'(y_i) \right\rangle_{\Gamma'} \geqslant 0,$$

and in $\Gamma' = F^+_{\alpha_{p-1}} \cdots F^+_{\alpha_1}(\Gamma)$ the vertex β is $(+)$-admissible; hence in the vector $w'(y_i)$ the βth coordinate is nonpositive. But since $y = \Sigma_0^s \, y_i$, it follows that $w'(y) = \Sigma_0^s \, w'(y_i)$ is equal to one of the terms of the sequence Π, and $w'(y)$ has a negative βth coordinate. This contradicts the fact that y is the first term in Π with nonpositive coordinates.

Note that uniqueness of the decomposition follows from its construction.

It also follows from the proof that an indecomposable vector is a root. Conversely, any root is indecomposable. Indeed, if $x = a + b$, $a, b \in U^+$, $\langle a, b \rangle \geqslant 0$, $\langle b, a \rangle \geqslant 0$, and x is a root, then $1 = \langle x, x \rangle = \langle a, a \rangle + \langle b, b \rangle + \langle a, b \rangle + \langle b, a \rangle \geqslant 2$.

It follows easily from the definition of a representation of general position V_0 in dimension x that $| \operatorname{Hom}_\Lambda(V_0, V_0) | = \langle x, x \rangle$. By Theorem 1, for any representation V in dimension x we have

$$| \operatorname{Hom}_\Lambda(V, V) | \geqslant | \operatorname{Hom}_\Lambda(V, V) | - | \operatorname{Ext}^1_\Lambda(V, V) | = \langle x, x \rangle.$$

Suppose $x = \Sigma_1^t \, x_i$ is a canonical decomposition and V_i is the indecomposable representation corresponding to the root x_i. Then $V_0 = \oplus_1^t V_i$ is a representation of general position, since, by Proposition 1, $\operatorname{Ext}^1_\Lambda(V_i, V_j) = 0$ and therefore $| \operatorname{Hom}_\Lambda(V_0, V_0) | = \langle x, x \rangle$.

BIBLIOGRAPHY

1. I. N. Bernšteĭn, I. M. Gel'fand and V. A. Ponomarev, *Coxeter functors and Gabriel quivers*, Uspehi Mat. Nauk **28** (1973), no. 2(170), 19–33; English transl. in Russian Math. Surveys **28** (1973).

2. Peter Gabriel, *Unzerlegbare Darstellungen*. I, Manuscripta Math. **6** (1972), 71–103, 309.

3. S. A. Krugljak, *Representations of algebras for which the square of the radical equals zero*, Zap. Naučn. Sem. Leningrad. Otdel. Mat. Inst. Steklov. (LOMI) **28** (1972), 60–68; English transl. in J. Soviet Math. **3** (1975), no. 5.

4. Sheila Brenner, *Quivers with commutativity conditions and some phenomenology of forms*, Proc. Internat. Conf. Representations of Algebras (Ottawa, 1974), Paper no. 5, Carleton Math. Lecture Notes, No. 9, Carleton Univ., Ottawa, Ont., 1974; reprint, Lecture Notes in Math., Vol. 488, Springer-Verlag, 1975, pp. 29–53.

5. Maurice Auslander and Idun Peiten, *Representation theory of Artin algebras*. III, Comm. Algebra **3** (1975), 239–294.

6. A. V. Roĭter, *Roots of integral quadratic forms*, Trudy Mat. Inst. Steklov. **148** (1978), 201–210; English transl. in Proc. Steklov Inst. Math. **1980**, no. 4(148).

7. Claus Michael Ringel, *Representations of K-species and bimodules*, J. Algebra **41** (1976), 269–302.

8. Ju. A. Drozd, *Coxeter transformations and representations of partially ordered sets*, Funkcional. Anal. i Priložen. **8** (1974), no. 3, 34–42; English transl. in Functional Anal. Appl. **8** (1974).

9. Ju. V. Roganov, *The dimension of the tensor algebra of a projective bimodule*, Mat. Zametki **18** (1975), 895–902; English transl. in Math. Notes **18** (1975).

10. M. M. Kleĭner and A. V. Roĭter, *Representations of differential graded categories*, Matrix Problems, Inst. Mat. Akad. Nauk Ukrain. SSR, Kiev, 1977, pp. 5–70. (Russian)

Translated by G. A. KANDALL

Amer. Math. Soc. Transl.
(2) Vol. **119**, 1983

On Homomorphisms Onto Finite Groups*

A. I. MAL'CEV

We shall use the word *algebra* in the sense of *algebraic system*; that is, an arbitrary set A of elements equipped with some system of operations $\varphi_1, \ldots, \varphi_s$ indexed in a definite order and called *basic operations* of the algebra. An algebra A is called *residually finite* if for any two distinct elements a and b in A there is a homomorphism σ of A into a finite algebra under which the images a^σ and b^σ of these elements are distinct.

The concept of residual finiteness turned out to be important above all in the theory of totally disconnected compact topological groups (Pontrjagin [16]). On the other hand, the author proved in [8] that finitely generated matrix groups are residually finite, and therefore so also are arbitrary free groups and free products of residually finite groups. The residual finiteness of free groups was later shown independently by Iwasawa in [4]. Hirsch in [2] established the residual finiteness of soluble groups satisfying the maximal condition on subgroups. Finally Gruenberg in [3] recently demonstrated the residual finiteness of free soluble groups, of an important class of groups analogous to the free soluble groups, and also of certain kinds of factor groups of free products of residually finite groups.

If, instead of homomorphic mappings onto finite algebras, we consider homomorphisms into the algebras of some other class K, we arrive at the definition of residual K-ness. As an example of a residual-type theorem many mention the theorem of Magnus that free groups are residually nilpotent. The general question of the residual nilpotency of free products of groups and of reduced free algebras was studied in [10] and [12]. There are some remarks on the residual K-ness of groups for some other classes K in the aforementioned paper of Gruenberg.

By analogy with residual finiteness let us agree to say that a subalgebra B of an algebra A is *finitely separable from an element* $a \notin B$ if there exists a homomorphism τ of A into some finite algebra under which $a^\tau \notin B^\tau$. We shall say that an algebra A is *finitely separable* if every subalgebra of it is finitely separable from all

---

1980 *Mathematics Subject Classification.* Primary 08A30, 20E26; Secondary 20D99, 20M14, 20F16.
*Translation of Učen. Zap. Ivanov. Gos. Ped. Inst. **18** (1958), 49–60.

elements not contained in that subalgebra. Under fairly general assumptions it is easy to see that finite separability is a stronger property than residual finiteness. In this paper we shall study the structure finitely separable soluble groups in detail. In particular, it turns out that soluble groups with the ascending chain condition on subgroups are finitely separable, which generalizes Hirsch's theorem, mentioned above, that such groups are residually finite.

In the final section (§7) we shall show that the well-known algorithmic problem of deciding whether an element belongs to a subalgebra has a positive solution in the classes of finitely separable algebras. This proves that for nilpotent groups of a given nilpotency class there exists an algorithm to decide whether an element belongs to a subgroup or not.

In §1 some new cases of residual finiteness in groups and subgroups are found, and in §6 the property of finite separability is interpreted from the point of view of a topology which naturally turns up in residually finite algebras.

§1. We listed above well-known sufficient conditions for residual finiteness of different classes of groups. We append here a few remarks on the problem of finding classes of residually finite groups and subgroups.

First of all, the residual finiteness of abelian groups can be formulated as follows: *an abelian group G is residually finite if and only if it contains no complete elements*; that is, elements $a \neq 1$ for which the equation $x^n = a$ has solutions for every natural number n.

Indeed, if $x^n = a$ is not solvable in G then $a \notin G^n$, where G^n denotes the subgroup generated by all nth powers of elements of G. The elements of G/G^n have bounded order, and therefore G/G^n is a finite direct product of cyclic groups. But a direct product of cyclic groups is residually finite. Therefore there exists a normal N/G^n of finite index for which $a \notin N$; that is, G is residually finite.

The above condition is necessary for the residual finiteness of *arbitrary* groups. However, the example to be constructed in §4 shows that this condition is not sufficient even for metabelian groups.

Fairly general sufficient conditions for the residual finiteness of soluble groups of finite reduced rank will be given later.

By an argument similar to that of P. Hall in [3] we obtain

THEOREM 1. *A semidirect product of a residually finite subgroup by a finitely generated residually finite normal subgroup is residually finite.*

Recall that a group G is the semidirect product of a normal subgroup H and a subgroup D if $G = DH$ and $D \cap H = 1$. The separability of an element $a \in G$, $a \neq 1$, from 1 follows directly from the hypothesis for $a \notin H$. Suppose $a \in H$. By hypothesis there exists in H a normal subgroup M of finite index that does not contain a. We denote by $f_\alpha(x_1,\ldots,x_n) = 1$ all the identities occurring in the factor group H/M, where n is the order of H/M; and we let N be the subgroup of H generated by elements of the form $f_\alpha(h_1,\ldots,h_n)$, where $h_1,\ldots,h_n$ run independently through H. According to B. Neumann, N has finite index in H. Moreover, N is a normal subgroup of the original group G. Since $N \subseteq M$, it follows that $a \notin N$. We consider the factor group $G = G/N$ and the images D_1 and H_1 of D and H in it. We have $G_1 = D_1 H_1$, $D_1 \cap H_1 = 1$, where H_1 is a finite normal subgroup of G_1. We denote by R_1 the set of elements of D_1 that commute with each element of H_1. Since H_1 is finite, R_1 is a normal subgroup of finite index in D_1. Hence $R_1 H_1$ has finite index in G_1. Since $[R_1, H_1] = 1$ and $R_1 \cap H_1 = 1$, it follows that $R_1 H_1$ is the direct product of the residually finite groups R_1 and H_1 and is therefore residually finite. Therefore G_1, which is a finite extension of $R_1 H_1$, is also residually finite, as required.

It follows from Theorem 1, for example, that groups that have a finite normal series in which each of the factor groups is either finite, infinite cyclic or finitely generated free groups of finite rank, are residually finite.

THEOREM 2. *Finitely generated commutative semigroups are residually finite.*

We consider the linear semigroup algebra for the given semigroup, say, over the rational numbers. By [9] the algebra admits an isomorphic matrix representation. Therefore the original semigroup turns out to have an isomorphic matrix representation over some field. It was proved in [8] that every finitely generated matrix group is residually finite. The same argument yields the residual finiteness of finitely generated semigroups of matrices. Thus the given semigroup is residually finite.

Finally, we consider any primitive class L of nonassociative rings characterized by some system of *multilinear* identities. It was proved in [12] that the free linear algebras in classes characterized by multilinear identities are residually nilpotent. The proof of this proposition goes over to rings. Therefore the free L-rings are residually nilpotent.

On the other hand, we shall show in §2 that all nilpotent rings which have a finite system of generators are residually finite. We conclude that *in the classes of rings characterized by multilinear identities the free rings are residually finite.*

In particular, free associative rings, free Lie rings, free alternative rings, etc., are residually finte.

§2. Suppose that K is a class of abstract algebras. We shall say that A is an algebra with *K-separable subalgebras* if for every $a \in A$ and every subalgebra B of A not containing a there exists a homomorphism σ of A into a suitable K-algebra for which $a^\sigma \notin B^\sigma$. If K is the aggregate of all finite algebras, then K-separability will be called finite separability. In the case of groups and rings K-separability implies residual K-ness.

It is clear that all subalgebras of an algebra with K-separable subalgebras are themselves algebras with K-separable subalgebras. A little less evident is

THEOREM 3. *Suppose that the class K contains all homomorphic images of subalgebras of algebras in K and that an algebra A has K-separable subalgebras.*

Then, if congruences on A commute, all homomorphic images of A are algebras with K-separable subalgebras.

Suppose that we have a homomorphism σ of A onto A_1 and suppose that B_1 is some subalgebra of A_1 and that $a_1 \in A_1$ but $a_1 \notin B_1$. We denote by B the full preimage of B_1 in A and let θ be the congruence on A corresponding to σ, $a^\sigma = a_1$. By hypothesis there exists a congruence η on A such that A/η is isomorphic to some subalgebra of a K-algebra and for any $b \in B$ we have $a\bar\eta b$. Since congruences commute by hypothesis it follows that the product $\eta\theta$ is a congruence on A (see [1]). The mapping $A \to A/\eta\theta$ is a homomorphism into an appropriate K-algebra, and at the same time $a^{\eta\theta}$ does not belong to $B^{\eta\theta}$. In fact, the contrary would mean that for some $b \in B$ we have $a(\eta\theta)b$; that is, for some $z \in A$ we have $a\eta z$ and $z\theta b$. Since B is the full preimage of B_1, it follows from $z\theta b$, $b \in B$, that $z \in B$, and the relation $a\eta z$ now contradicts the assumption about θ.

It follows at once from Theorem 3 that if there exist noncyclic simple groups which do not belong to the class K, then free groups with a sufficient number of free generators contain subgroups which are not K-separable. In fact, every group is a factor group of a free group. If free groups had separable subgroups, then so would all of their factor groups.

On the other hand, all subgroups of a cyclic group are finitely separable, and at the same time any countable abelian group is a factor group of a direct product of a countable number of cyclic groups. Among the countable abelian groups there are groups which have no nontrivial subgroups of finite index. It is trivial that these groups are not groups with finitely separable subgroups. Therefore *a direct product of a countable number of groups with finitely separable subgroups is not necessarily a group with finitely separable subgroups.* However, for direct products of a *finite* number of groups the situation is more simple.

THEOREM 4. *If K is a class of groups or rings that contains direct products of arbitrary pairs of algebras in it, then a direct product of a finite number of algebras with K-separable subalgebras is an algebra with K-separable subalgebras.*

It is sufficient to consider a direct product of two groups $A = F \times G$. Suppose that B is some subgroup of A; C and D are canonical images of B in F and G respectively, $B \cap F = M$ and $a \in A$, $a \notin B$, $a = cd$, $c \in F$, $d \in G$. If $a \notin CD$, then $c \notin C$ or $d \notin D$ and in order to prove the result it is enough to consider the groups A/F and A/G. We therefore assume that $a \in CD$. Let $c_0 d \in B$ and $c_0 \in C$. By hypothesis the element cc_0^{-1} and the subgroup M which does not contain it are separable in F. This means that there exists a normal subgroup F_0 of F such that F/F_0 is isomorphic to a subgroup of a K-group and $cc_0^{-1} \notin F_0 M$. We shall show that $a \notin F_0 B$. Suppose that, on the contrary, $a = fc_1 d_1$, $f \in F_0$, $c_1 d_1 \in B$, whence $d = d_1$ and $c = fc_1$. Since $c_1 d \in B$ and $c_0 d \in B$, it follows that $c_1 c_0^{-1} \in B \cap F = M$. Hence $cc_0^{-1} = fc_1 c_0^{-1} \in F_0 M$, contrary to the choice of M. Thus the image of a in the canonical mapping of $A = F \times G$ onto $F/F_0 \times G$ is not in the image of B. Applying our argument once again to the product $F/F_0 \times G$, we obtain that the image of a in $F/F_0 \times G/G_0$ does not belong to the image of B in $F/F_0 \times G/G_0$, and this group is isomorphic to a subgroup of a suitable K-group, as required. The proof for rings is similar.

§3. We want now to find conditions under which soluble groups have finitely separable subgroups. We begin with abelian groups. As a preliminary we note that *an abelian group is a group with finitely separable subgroups if and only if its factor groups are all residually finite.* The necessity part follows from Theorem 3, and the sufficiency is clear.

A periodic abelian group has finitely many separable subgroups if and only if the orders of the elements of its primary components are bounded.

The sufficiency follows directly from the fact that the groups of the type mentioned together with all of their factor groups decompose into direct products of cyclic groups and are therefore residually finite.

In order to prove necessity we assume that the abelian group A contains elements $a_1, a_2, \ldots$ of orders $p, p^2, \ldots$ respectively. The subgroup B generated by $a_1, a_2, \ldots$ is countable. We therefore have two possible cases:

1) B has an element of infinite height; that is, an element $b \neq 1$ for which the equation $x^{p^n} = b$ is solvable for any natural number n.

2) B decomposes into a direct sum of cyclic groups whose orders are not uniformly bounded.

In the first case B is not residually finite, since if C were a subgroup of finite index and $b \notin C$ then the group B/C would be a finite p-group in which $b \neq 1$ and the equation $x^{p^n} = b$ is solvable for all n, which is impossible. In the second case B can be mapped homomorphically onto an infinite quasicyclic p-group P. Suppose $p \cong B/D$. Then A/D contains an infinite quasicyclic p-group and so cannot be residually finite. Thus in neither case is A a group with finitely separable subgroups.

We shall say that a torsion free abelian group is *bounded* if it has finite rank and if for any element a and any subgroup B not containing a the congruence $x^{p^n} \equiv a(B)$ for every fixed p is solvable only for finitely many nonnegative n.

It is easy to see that a torsion free abelian group of rank 1 is bounded if and only if its characteristics (see [5], Chapter VIII, §30) do not contain the symbol ∞.

In the general case a torsion free abelian group is bounded if and only if it is isomorphic to a subgroup of a direct product of bounded groups of rank 1.

A torsion free abelian group has finitely separable subgroups if and only if it is bounded.

Indeed, an abelian group of infinite rank contains free abelian subgroups of infinite rank and therefore cannot be a group with finitely separable subgroups. Suppose that A has rank r and $a_1, \ldots, a_r$ are arbitrary linearly independent elements of A. Then in additive notation every element of A can be represented in the form

$$a = \frac{m_1}{n_1} a_1 + \cdots + \frac{m_r}{n_r} a_r \qquad ((m_i, n_i) = 1; i = 1, \ldots, r). \tag{1}$$

If for all a the denominator n_1 contains some prime p to arbitrarily large powers, then the factor group of A modulo the subgroup N generated by $a_1, \ldots, a_r$ contains a quasicyclic p-group, and so A/N cannot be residually finite. Thus, if A has finitely separable subgroups, then the denominators $n_1, \ldots, n_r$ in (1) can only be divisible by bounded powers of prime numbers, and hence A is bounded.

Conversely, suppose that A is bounded. From (1) it easily follows that every factor group A_1 of A has the following structure: the periodic part $F \subset A_1$ has finite primary components, and A_1/F is bounded. On the other hand, it is well known that for any abelian group A and any element a of A the factor group A/M, where M is a maximal subgroup not containing a, is either finite or an infinite quasicyclic p-group. Since the latter case is impossible for bounded groups, it follows that A_1 is residually finite.

THEOREM 5. *An abelian group A has finitely separable subgroups if and only if the factor group of A modulo its periodic part F is bounded and the orders of elements of each primary component of F are uniformly bounded.*

The necessity follows from the preceding results, since if A is a group with separable subgroups then both F and A/F must both have the same property.

On the other hand, factor groups of groups having the structure mentioned in Theorem 5 have the same structure and therefore cannot be quasicyclic p-groups. In view of earlier remarks this implies the residual finiteness of all factor groups, as required.

§4. We consider the following example. Suppose that the group G is given by generators a_i, b_i and z $(i = 1, 2, \ldots)$ and defining relations

$$a_i^2 = b_i^2 = 1, \qquad a_i b_i = b_i a_i z,$$

$$a_i a_j = a_j a_i, \qquad a_i b_j = b_j a_i, \qquad b_i b_j = b_j b_i \qquad (i \neq j; \ i, j = 1, 2, \ldots).$$

This group has commutator subgroup $H = \{1, z\}$, which coincides with the center of G. The central factor group is a finite direct sum of an infinite number of cyclic groups of order 2. Thus G/H and H are abelian groups with finitely separable subgroups. However, G is not residually finite. In fact, suppose that N is a normal subgroup of finite index in G. Then for suitable $i \neq j$ we have $a_i \equiv a_j \ (N)$ or $a_i a_j \in N$. Therefore $b_i^{-1} a_i a_j b_i = a_i a_j z \in N$, $z \in N$; that is, all normal subgroups of finite index contain z, and therefore z is not finitely separable from the identity subgroup.

This example shows that in the study of conditions which one must impose on the factor groups of soluble groups in order that they later have finitely separable subgroups the necessary and sufficient conditions mentioned earlier are too broad. We therefore introduce the following definition: an abelian group G will be called *bounded* if all of the primary components of its periodic part F are finite and the factor group G/F is a bounded torsion free group in the sense defined earlier.

Thus the new definition of boundedness coincides with the old for torsion free abelian groups, and imposes more stringent restrictions only on the periodic parts of the groups.

Subgroups, factor groups and direct products of a finite number of bounded abelian groups are bounded abelian groups.

The assertions are clear for subgroups and direct products. It is immediately verifiable that all true factor groups of a torsion free bounded group of rank 1 are periodic groups with cyclic primary components.

Suppose now that A is a torsion free bounded group of rank r and that B is a nontrivial subgroup of A. We choose in B a maximal linearly independent set of elements $a_1,\ldots,a_s$ and add to it elements $a_{s+1},\ldots,a_r$ to form a basis for A. All elements of A can be uniquely represented by linear forms (1) in the elements $a_1,\ldots,a_r$ with rational coefficients. Suppose G_i is the group of forms $m_i a_i$, where the coefficients are chosen as they occur in n_i decompositions of elements of A. It is clear that the G_i $(i = 1,\ldots,r)$ are bounded torsion free groups of rank 1 and A is a subgroup of the direct sum $G = G_1 + \cdots + G_r$. We denote by B_0 the subgroup of B generated by the elements $a_1,\ldots,a_s$, and we set $H = G_1 + \cdots + G_s$. Then $A/B \subset G/B \cong H/B + A_1$, where $A_1 = G_{s+1} + \cdots + G_r$ and $H/B_0/B/B_0$; H/B_0 is a periodic bounded group and hence so also is H/B. This proves that factor groups of torsion free bounded groups are bounded. It is easy to reduce the case of mixed bounded groups to that of torsion free groups.

§5. A soluble group G is said to be *bounded* if it has a finite normal series with bounded abelian factors.

It follows at once from Schreier's refinement theorem and the preceding results that *the factors of any soluble finite normal series of a bounded soluble group are bounded abelian groups.* From this it follows in turn that *subgroups, factor groups and direct products of finitely bounded soluble groups are bounded soluble groups.*

Since a bounded soluble group possesses a finite normal series whose factors are abelian groups of finite reduced rank, every bounded soluble group is an A_1-group in the terminology of [13]. According to [13] the factor group of a soluble A_1-group modulo its maximal normal periodic subgroup F is an A_4-group; that is, G/F has a finite normal series whose factors are abelian groups with finite periodic parts.

We recall that the *special reduced rank* of a group G is the least r such that any finite set of elements generates an r-generator subgroup of G. Soluble A_4-groups have finite special rank, and hence the factor group of a bounded soluble group modulo its maximal periodic normal subgroup has finite special rank.

LEMMA 1. *Let G^n denote the subgroup of the group G generated by the nth powers of elements of G. If G is a bounded soluble group, then G/G^n $(n = 1, 2,\ldots)$ is finite.*

In fact, G/G^n is a bounded periodic soluble group, and the orders of its elements divide n. It follows that the factors of a soluble normal series for G/G^n

have only finitely many primary components and hence are finite, so that G/G^n is also finite.

LEMMA 2. *If G is a nilpotent group with lower central series of length n, then the equation $x^m = a$ is solvable in G^{m^n} ($m, n = 1, 2, \ldots$) for any $a \in G^{m^n}$ (cf. [11]).*

Suppose that $G \supset G_1 \supset \cdots \supset G_n \supset 1$ is the lower central series of a free nilpotent group of class n with free generators $x_1, \ldots, x_s$, and that the lemma is true for groups of lower class. Then in G/G_n there is an identity of the form

$$x_1^{m^{n-1}} x_2^{m^{n-1}} \cdots x_s^{m^{n-1}} = \left(x_{i_1} \cdots x_{i_t} \right)^m = \left(\pi(x) \right)^m.$$

It follows from this that

$$x_1^{m^{n-1}} x_2^{m^{n-1}} \cdots x_s^{m^{n-1}} = \pi((x))^m u, \tag{2}$$

in G, where $u \in G_n$. Suppose that $v_1, \ldots, v_p$ are generators of the factor group G_{n-1}/G_n. The commutators $[v_\alpha, x_\beta]$ will generate G_n, and therefore

$$u = \Pi \left(v_{\alpha_i}, x_{\beta_i} \right)^{\varepsilon_i}.$$

Substituting in (2) the expression x_i^m in place of x_i and noting that the $[v_\alpha, x_\beta]$ belong to the center of G and $[v_\alpha, xy] = [v_\alpha, x][v_\alpha, y]$, we obtain

$$x_1^{m^n} \cdots x_s^{m^n} = \left(\pi(x^m) \right)^m \Pi \left(\bar{v}_{\alpha_i}, x_{\beta_i} \right)^{m \varepsilon_i} = \left[\pi(x^m) \Pi \left(\bar{v}_{\alpha_i}, x_{\beta_i} \right)^{\varepsilon_i} \right]^m,$$

as required.

LEMMA 3. *All bounded soluble groups are residually finite.*

This was established for abelian groups in §3. We therefore proceed on the inductive assumption that G has a soluble normal series $G \supset G_1 \supset \cdots \supset G_n \supset 1$ of length n and that the lemma is true for groups with soluble series of length less than n.

Suppose that a is any nonidentity element of G. If $a \notin G_1$, then the image $\bar{a}$ of a in G/G_1 is different from 1. Since G/G_1 is abelian, there is a normal subgroup N/G_1 of finite index which does not contain a, as required. Suppose, then, that $a \in G_1$. By hypothesis there is a normal subgroup M of finite index s in G_1 which does not contain a. By Lemma 1, G_1^s has finite index in G_1 and is normal in G, and $G_1^s \subseteq M$. Let $H = G/G_1^s$, and let F and b be the images of G_1 and a in H, respectively. We have that $b \neq 1$, $b \in F$, F is a finite normal subgroup of H and H/F is abelian. We denote by T the set of all elements of H which commute with each element of F. Since F is finite, the index of T in H is also finite. If $b \notin T$, then T will do as the required normal subgroup. Suppose that $b \in T$, and set $T \cap F = T_0$. Then T_0 is a finite central subgroup of T, and T/T_0 is abelian. Thus T is nilpotent of class 2. Suppose that the order of T_0 is m. The subgroup T^{m^2} has finite index in T, and will therefore do as the subgroup we need if $b \notin T^{m^2}$. Assume that $b \in T^{m^2}$. By Lemma 2 there are elements $b_1, b_2, \ldots$ in T^{m^2} satisfying the relations $b_1^m = b$, $b_2^m = b_1, \ldots$. These elements generate an abelian subgroup which contains an infinite quasicyclic p-group, where p is a prime divisor of m. But this is a contradiction, since T and all of its subgroups are bounded.

In particular, it follows from Lemma 3 that every finite subgroup H of a bounded soluble group G is finitely separable.

In fact, suppose that $a \notin H$, $H = \{h_1, \ldots, h_r\}$. By Lemma 3 there are normal subgroups N_i of finite index in G such that $ah_i^{-1} \notin N_i$ $(i = 1, \ldots, r)$. Then $N = \cap\, N_i$ is the required normal subgroup of finite index such that $a \notin HN$.

THEOREM 6. *All subgroups of bounded soluble groups are finitely separable.*

This result was proved for abelian groups in §3. We may therefore assume by induction that the group G in question has a soluble normal series $G \supset G_1 \supset \cdots \supset G_n \supset 1$ of length n, and that the theorem is proved for those with shorter length.

Suppose that B is a subgroup of G, and that $a \in G$ but $a \notin B$. If $a \notin BG_1$, then the abelian group G/BG_1 has a subgroup N/BG_1 of finite index which does not contain a. Then N will do as the required normal subgroup separating a from B.

Assume that $a \in BG_1$. Let $a = bh$, $b \in B$, $h \in G_1$, and $T = B \cap G_1$. Since $h \notin T$ and G_1 has a soluble normal series of length $n - 1$, there is a normal subgroup M of finite index in G_1 for which $h \notin TM$. Let s denote the index of M in G_1. Then G_1/G_1^s is finite, G_1^s is normal in G and $h^s \notin TG_1^s$. If there were $a \in BG_1^s$, whence $a = b_1 h_1$ $(b_1 \in B, h_1 \in G_1^s)$, then we would have $bh = b_1 h_1$, $b^{-1}b_1 = hh_1^{-1} \in B \cap G_1 = T$, $h \in TG_1^s$, which contradicts the hypothesis. Thus $a \notin BG_1^s$.

Set $G/G_1^s = H$ and suppose that the images of G_1, B and a in H are F, C and b, respectively. We have that F is a finite normal subgroup of H, H/F is abelian and $b \notin C$. Then H contains the normal subgroup FC in which C is a subgroup of finite index. Set $N = (FC)^t$, where t is the index of C in FC. Then N is a normal subgroup of H contained in C. We consider $H/N = H_1$ and let C_1 and b_1 be the images of C and b in H_1. Since $b_1 \notin C_1$ and C_1 is finite, by a corollary of Lemma 3 there is a normal subgroup N_1 of finite index in H_1 such that $b_1 \notin C_1 N_1$. We have constructed a chain of homomorphisms $G \to H \to G_1 \to H_1/N_1$. Putting them together yields a homomorphism of G onto the finite group H_1/N_1 such that the image of a does not belong to the image of B, as required.

The example of abelian groups studied in §3 shows that boundedness of soluble groups is not necessary for finite separability of their subgroups. However, it is evident from the foregoing arguments that *boundedness is not only sufficient but also necessary to ensure finite separability of subgroups in torsion free soluble groups.*

In fact, if subgroups of a soluble group G are separable it follows also that subgroups of factor groups of subgroups of G are separable and hence subgroups of factors of a soluble normal series for G are separable. The reduced ranks of these factors are therefore finite, and G is an A_1-group. But a torsion free A_1-group is an A_4-group, and so G is bounded.

We remark at this point that *if the additive group of a ring R is bounded, then R has separable subrings.*

In fact, it was shown earlier that the normal separating subgroups may be taken to be the subgroups of the form nR, which in the case of a ring are automatically ideals.

In particular, subrings of finitely generated nilpotent rings are finitely separable since additive groups of such rings are finitely generated.

§6. Let A be an arbitrary algebra and $S = \{\theta_\alpha\}$ a system of congruences on A. We assume that for any two congruences θ_α and θ_β in S there is a $\theta_\gamma \in S$ such that $\theta_\gamma \leq \theta_\alpha \cap \theta_\beta$, and that the intersection of all congruences in S is zero. Following Hämisch [17], we call the residue classes modulo congruences in S, and also arbitrary unions of residue classes, *S-open sets*. By the hypotheses on S it follows that the S-open sets define a Hausdorff topology on A in which the basic operations of A are continuous. Thus the abstract algebra A is turned into a topological algebra whose topology is called the *S-topology*.

Suppose now that we are given an abstract algebra A and some class of algebras K of the same type which is closed under taking direct products of pairs of algebras. Each homomorphism of A into an arbitrary K-algebra defines a congruence on A. Let S be the collection of all congruences on A obtained in this way. If A is residually a K-algebra, then the intersection of all congruences in S is 0 and we can talk about the S-topology on A, which we shall also call the *K-topology*. It is clear that every K-topology is totally disconnected (see [17]) and that any homomorphism of A with its K-topology into an arbitrary K-algebra, viewed as a topological algebra with the discrete topology, is continuous and open.

Suppose that A contains a subalgebra B and that $a \in A$ but $a \notin B$. We assume for a homomorphism σ of A into some K-algebra that $a^\sigma \notin B^\sigma$. Then the residue class $[a]$ modulo the congruence θ corresponding to σ that contains a, does not intersect B. Since $[a]$ is an open set, we see that the K-separability of B from a is equivalent to the open separability of a from B. In particular, the K-separability of B from all the elements of A not contained in B turns out to be equivalent to the closedness of B in the K-topology. Hence *the algebras with K-separable subalgebras are precisely those algebras whose abstract subalgebras are closed in the K-topology.*

Let us agree to call the K-topology *finite* if K is the class of all finite algebras. The problem considered in §5 can now be reformulated as the problem of investigating those soluble groups all subgroups of which turn out to be closed in the finite topology.

§7. We consider finally the connection between residual finiteness, finite separability and some algorithmic questions. The fundamental work in this area is that of McKinsey [7].

Suppose that K is a class of abstract algebras whose basic operations are denoted by the symbols $f_1, \ldots, f_s$. Instead of them we introduce predicate symbols $P_1, \ldots, P_s$, regarding $P_i(x_1, \ldots, x_{n_i})$ as equivalent to the assertion $x_{n_i} = f_i(x_1, \ldots, x_{n_{i-1}})$. With the help of the symbols P_i and the object variables $x_1, x_2, \ldots$

we can compose the formulas of the restricted predicate calculus (cf. [7] and [14]). According to McKinsey the system Σ of axioms is called *finitely reducible* with respect to a class of algebras K, if for every $\alpha \in \Sigma$ which does not hold in some K-algebra there is a finite K-algebra in which α does not hold.

The following theorem of McKinsey holds: *Suppose that the class Σ of axioms is reducible with respect to a finitely axiomatizable class of algebras K. Then there exists an algorithm to decide whether an arbitrary fixed axiom of Σ is true or not true for all K-algebras* [7].

We assume that the class K of algebras is restrictively primitive (see [14]) and let A be an algebra in this class given by a finite number of generators $a_1, \ldots, a_n$ and defining equalities $h_1 = g_1, \ldots, h_r = g_r$, where the h_λ and g_λ are polynomials in $a_1, \ldots, a_n$. The identity problem for K consists in finding an algorithm to decide the truth of any equality of the form $h = g$ in A, where g and h are polynomials. Since the truth of the stated equality in A is equivalent to the truth of the conditional equality $h_1 = g_1 \,\&\, \cdots \,\&\, h_r = g_r \to f = g$ in all algebras of the class K, it follows from McKinsey's theorem (cf. Kuznecov [6]) that if all algebras in K are residually finite then the identity problem for algebras in K given by defining relations is algorithmically solvable. Taking Theorem 2 into account, we see, in particular, that the word problem for commutative semigroups is algorithmically solvable. This result was obtained independently by V. A. Emeličev and G. S. Ceĭtin using another method.

Suppose now that K is again a class of primitive algebras and that A is an algebra in this class given by a finite number of generators and defining relations. We assume in addition that we are given finitely many elements of A represented in the form of polynomials $h, h_1, \ldots, h_s$ in the generators $a_1, \ldots, a_n$ of A. We want to find an algorithm that will tell us whether h belongs to the subalgebra B generated by $h_1, \ldots, h_s$. We shall call this the problem of the occurrence of an element in a subalgebra.

If an algebra A from the primitive class K, given by finitely many generators and defining relations, is residually finite and all of its subalgebras are finitely separable, then the problem of the occurrence of an element in a subalgebra of A is algorithmically solvable.

In order to prove this, similarly to the method of McKinsey, we arrange all finite algebras of the given type in a sequence $A_1, A_2, \ldots$, and from each of these we choose, in all possible ways, as many elements $a_i^m \in A_m$ $(i = 1, \ldots, n)$ as there are generators $a_1, \ldots, a_n$ in A. We assume that $h, h_1, \ldots, h_s$ are fixed polynomials in $a_1, \ldots, a_n$, and we arrange in a sequence $F_1, F_2, \ldots$ all the polynomials in $h_1, \ldots, h_s$. The required algorithm Φ is described as follows: the $(2m - 1)$st step consists in taking the algebra A_m and verifying whether $a_1^m, \ldots, a_n^m$ generate A_m and whether they satisfy the defining relations for A. If not, go to the next step of Φ. If yes, then decide whether the element $h(a_1^m, \ldots, a_n^m)$ occurs in the subalgebra of A_m generated by the elements $h_i(a_1^m, \ldots, a_n^m)$. If not, then the process terminates and the element h in A does not belong to the subalgebra generated by the elements $h_1, \ldots, h_s$. If yes, then go to step $2m$, which consists in deciding whether the equality $h = F_m$ holds in A. The algorithm for this is indicated by McKinsey's theorem. If $h = F_m$, then the process terminates and h belongs to the subalgebra generated by $h_1, \ldots, h_s$. If $h \neq F_m$, then we go to the next step. The finite separability of the subalgebras of A guarantees that the process Φ terminates.

It now follows from Theorem 6 that *the problem of the occurrence of an element in a subalgebra is algorithmically solvable for the class of all nilpotent groups of class k.*

§8. More general still is the so-called *intersection problem* for subalgebras: in a primitive class K suppose we are given an algebra A by means of generators and finitely many defining equalities. Suppose also that we are given two systems of polynomials $g_1, \ldots, g_m$ and $h_1, \ldots, h_n$ in the generators of A. We are required to find generators for the intersection of the subalgebras of A generated by $g_1, \ldots, g_m$ and $h_1, \ldots, h_n$, respectively. Instead of finding generators we can pose more special questions as to whether the intersection is the identity, or whether it is nonempty. The author does not know whether one can link the intersection problem with some sort of "separability" by McKinsey's method. However, using the method of our note [15], one can easily find an algorithm to solve the intersection problem for subgroups in the class of nilpotent groups of arbitrary fixed nilpotency class.

It is more difficult to treat the isomorphism problem for groups of fixed nilpotency class given by generators and relations, although it is possible to link this problem with the following finite separability.

Suppose that the algebra A belongs to a quasiprimitive class K which contains a quasiprimitive subclass L. We say that an algebra B is a *replica* of A in L if it has as generators all elements of A and the defining equalities in L are all equalities of the form $a = f_\nu(a_1, \ldots, a_n)$ which hold in A (the f_ν are symbols of the basic operations in K). The class L is said to be *locally fintie* if all of its finitely generated algebras are finite.

It is easy to prove that the isomorphism problem for the class K is certainly algorithmically solvable if for each pair A_1, A_2 of nonisomorphic finitely generated K-algebras there exists a quasiprimitive locally finite subclass $L \subseteq K$ in which the replicas of A_1 and A_2 are nonisomorphic.

The author does not know whether the class of nilpotent groups of fixed nilpotency class $k > 1$ has the above property. Abelian groups do have the property ([16], Chapter 1, §6, Proof of Theorem 2).

Analogously, the solution of the algorithmic conjugacy problem for nilpotent groups would be affirmative if for every finitely generated nilpotent group G and every pair of nonconjugate elements of G there exists a normal subgroup N of finite index in G such that the images of a and b in G/N are nonconjugate. Direct calculations show that free metabelian groups certainly possess this property.

BIBLIOGRAPHY

1. Garret Birkhoff, *Lattice theory*, 2nd rev. ed., Amer. Math. Soc., Providence, R.I., 1948.

2. K. A. Hirsch, *On infinite soluble groups*. IV, J. London Math. Soc. **27** (1952), 81–85.

3. K. W. Greenberg, *Residual properties of infinite soluble groups*, Proc. London Math. Soc. (3) **7** (1957), 29–62.

4. Kenkiti Iwasawa, *Einige Sätze über freie Gruppen*, Proc. Imp. Acad. Tokyo **19** (1943), 272–274.

5. A. G. Kuroš, *Theory of groups*, 2nd ed., GITTL, Moscow, 1953; English transl., Vols. I, II, Chelsea, New York, 1955, 1956, 1960.

6. A. V. Kuznecov, *On the problems of identity and functional completeness for algebraic systems*, Proc. Third All-Union Math. Congr. (Moscow, 1956), Vol. 2, Izdat. Akad. Nauk SSSR, Moscow, 1958, pp. 145–146. (Russian)

7. J. C. C. McKinsey, *The decision problem for some classes of sentences without quantifiers*, J. Symbolic Logic **8** (1943), 61–76.

8. A. I. Mal'cev, *On the faithful representation of infinite groups by matrices*, Mat. Sb. **8(50)** (1940), 405–422; English transl. in Amer. Math. Soc. Transl. (2) **45** (1965).

9. ______, *On the representations of infinite algebras*, Mat. Sb. **13(55)** (1943), 263–286. (Russian)

10. ______, *Generalized nilpotent algebras and their adjoint groups*, Mat. Sb. **25(67)** (1949), 347–366; English transl. in Amer. Math. Soc. Transl. (2) **69** (1968).

11. ______, *Nilpotent torsion-free groups*, Izv. Akad. Nauk SSSR Ser. Mat. **13** (1949), 201–212. (Russian)

12. ______, *On algebras defined by identities*, Mat. Sb. **26(68)** (1950), 19–33. (Russian)

13. ______, *On some classes of infinite solvable groups*, Mat. Sb. **28(70)** (1951), 567–588; English transl. in Amer. Math. Soc. Transl. (2) **2** (1956).

14. ______, *Quasiprimitive classes of abstract algebras*, Dokl., Akad. Nauk SSSR **108** (1956), 187–189; English transl., Chapter IV in A. I. Mal'cev, *The metamathematics of algebraic systems. Collected papers*: 1936–1967, North-Holland, 1971.

15. ______, *Two remarks on nilpotent groups*, Mat. Sb. **37(79)** (1955), 567–572. (Russian)

16. L. S. Pontrjagin, *Continuous groups*, 2nd ed., GITTL, Moscow, 1954; English transl., *Toplogical groups*, Gordon & Breach, New York, 1966.

17. Werner Hämisch, *Über die Topologie in der Algebra*, Math. Z. **60** (1954), 458–487.

Translated by J. C. LENNOX

Amer. Math. Soc. Transl.
(2) Vol. **119**, 1983

On Some Infinite Systems of Identities*

A. JU. OL'ŠANSKIĬ

Soon after it was discovered that not every variety of groups can be defined by a finite set of identities([1]) (see, for example, [2] and [3]), A. Tarski posed the following question: Can every variety of groups be defined by an irreducible (independent) system of identities? (See [4], 4.64.) In some cases this question can be answered affirmatively, as can be seen from Theorem 1 of the present note. Moreover, here we touch upon a question that is dual in some sense; the question of irredundant decompositions of varieties into a join. We will speak of systems of identities, verbal subgroups of a free group of countable rank, and varieties, without mentioning each time the obvious correspondences among these concepts.

A system of verbal subgroups $\{V_\alpha\}_{\alpha \in M}$ of a free group F of countable rank is called *independent* if $V_\alpha \not\subseteq \Pi_{\beta \in M \setminus \alpha} V_\beta$ for each $\alpha \in M$. A set of identities $\{f_i(x_1, \ldots, x_{n_i}) \equiv 1\}_{i \in M}$ is *independent* (i.e. none of these identities follows from the others) if the system of verbal subgroups corresponding to these identities is independent. A system of verbal subgroups $\{V_i\}_{i \in M}$ is *irredundant* if $V_i \not\supseteq \bigcap_{\alpha \in M \setminus i} V_\alpha$ for each $i \in M$.

Using the fact that there exists an infinite series of groups $G_1, G_2, \ldots$ (all G_i can be taken from a solvable, locally finite variety) such that

$$G_i \notin \mathrm{var}(G_j, j \neq i) \tag{1}$$

(see [2] and [5]), we obtain examples of an infinite irredundant system of varieties $\{\mathrm{var}(G_i)\}_{i=1,2,\ldots}$ and an infinite independent system $\{\mathrm{var}(G_j, j \neq i)\}_{i=1,2,\ldots}$. In general, these two concepts are related in the following way.

---

*Translation of Trudy Sem. Petrovsk. Vyp. 3 (1978), 139–146; MR **58**#11125.
1980 *Mathematics Subject Classification.* Primary 20E10.
([1]) The concepts and notation not defined here are explained in [1].

PROPOSITION 1. 1) *A system of verbal subgroups* $\{V_i\}_{i \in M}$ *is independent if and only if the system* $\{V_i^* = \prod_{\alpha \neq i} V_\alpha\}_{i \in M}$ *is irredundant. A system* $\{V_i\}_{i \in M}$ *is irredundant if and only if the system* $\{V_i^0 = \bigcap_{\alpha \neq i} V_\alpha\}_{i \in M}$ *is independent.*

2) *For an arbitrary system* $\{V_i\}_{i \in M}$ *we have the equalities* $V_i^* = V_i^{*0*}$, $V_i^0 = V_i^{0*0}$ *and the inclusions* $V_i^{0*} \subseteq V_i \subseteq V_i^{*0}$; *also,* $V_i^{*0} = V_i P$, *where* $P = \bigcap_{\alpha \in M} V_\alpha^*$, *and* $V_i^{0*} = V_i \cap S$, *where* $S = \prod_{\alpha \in M} V_\alpha^0$.

3) *If* $\{V_i\}_{i \in M}$ *is an independent system, then* $\prod_{i \in M}(V_i^{*0}/P)$ *is a direct product of the nonidentity groups* V_i^{*0}/P. *If* $\{V_i\}_{i \in M}$ *is an irredundant system and* $T = \bigcap_{i \in M} V_i^0$, *then* $\prod_{i \in M}(V_i^0/T)$ *is a direct product of the nonidentity groups* V_i^0/T.

We omit the proof of this proposition, since all of the assertions follow easily from the definitions.

With the aid of 3) (or directly using the existence of infinite independent systems) we can indicate various sublattices in the lattice of verbal subgroups of a free group. For example, if we index all subgroups of an infinite independent system by means of all rational numbers, i.e. $\{V_\alpha\}_{\alpha \in Q}$, then the verbal subgroups $U_\sigma = \prod_{\alpha \leq \sigma} V_\alpha$ (σ a real number) are obviously ordered by inclusion in the same way as the points of the real line.

This last observation and relation (1) give us ground to assert that there exists a continual set of locally finite varieties such that any two are related by means of inclusion. Therefore, the *order functions* $f_{\mathfrak{W}}$ of these varieties ($f_{\mathfrak{W}}(n)$ is the order of a free group of rank n in the variety $\mathfrak{W}$) are distinct; hence, in particular, Higman's "CREAM conjecture" is false.[*] (For details see [6], [1], and [7]. The proof of the continuality of the set of order functions, which is given in [7], is much more complicated.)

We now turn to the theorem; the possibility of relaxing the conditions and of generalizing the statement will be discussed below.

THEOREM 1. *Any solvable, locally finite variety* $\mathfrak{W}$ *can be defined by an irreducible system of identities.*

The following lemmas will enable us to prove this theorem.

LEMMA 1. *There exists a locally finite variety* $\mathfrak{U}$ *with a finite basis of identities such that* $\mathfrak{U} \supseteq \mathfrak{W}$.

PROOF. An exponent identity and an identity restricting the solvability length can obviously be used to define the variety $\mathfrak{U}$.

LEMMA 2. *Suppose a variety* $\mathfrak{U}$ *is generated by its finite groups and* $\mathfrak{W} \subsetneqq \mathfrak{U}$. *Then there exists a minimal variety* $\mathfrak{W}'$ *that strictly contains* $\mathfrak{W}$ *and lies in* $\mathfrak{U}$.

PROOF. It follows from the hypothesis of the lemma that there exists a finite group $G \in \mathfrak{U}$ such that $\text{var}(G) \not\subseteq \mathfrak{W}$. Since $\text{var}(G)$ is a Cross variety [1], it

---

[*]*Translator's note.* "CREAM" stands for "Closure under Repeated Exponentiation, Addition, and Multiplication".

contains some minimal subvariety $\mathfrak{M}$ such that $\mathfrak{M} \not\subseteq \mathfrak{W}$, but $\mathfrak{M}_1 \subseteq \mathfrak{W}$ for any $\mathfrak{M}_1 \subset \mathfrak{M}$. Put $\mathfrak{W}' = \mathfrak{W} \cup \mathfrak{M}$. Obviously, $\mathfrak{W} \subset \mathfrak{W}' \subseteq \mathfrak{U}$. Suppose $V' \subset W \subseteq V$ (we pass to verbal subgroups of a free group of countable rank). Then $W \not\subseteq M$ and $M_1 = WM \supset M$, i.e. $M_1 \supseteq V$ by the choice of the variety $\mathfrak{M}$. Now the modular law yields $W = V'W = (M \cap V)W = WM \cap V = M_1 \cap V = V$, as required.

REMARK 1. Since a free group F is residually finite, it follows from Lemma 2 that any nonidentity verbal subgroup V of F contains a maximal verbal (in F) subgroup V'.

LEMMA 3. *If $\mathfrak{W} = \mathfrak{W}_0 \subset \mathfrak{W}_1 \subset \cdots \subset \mathfrak{W}_n$ is a chain of varieties, each of which is maximal in the one following, $\mathfrak{W}_0 \subseteq \mathfrak{W} \subset \mathfrak{W}_n$, and $\mathfrak{U}$ is a finitely based variety, then $\mathfrak{W}$ is finitely based.*

PROOF. Each variety $\mathfrak{W}_{i-1}$, in view of maximality, is defined in $\mathfrak{W}_i$ by a single identity $v_i(x_1, \ldots, x_{n_i}) \equiv 1$. Thus, a basis of identities of the variety $\mathfrak{W}$ can be obtained by adjoining to a finite basis of $\mathfrak{U}$ all of the identities $v_i(x_1, \ldots, x_{n_i}) \equiv 1$, $i = 1, \ldots, n$.

LEMMA 4. *Suppose a collection of words $\mathfrak{F}_0 = \{ f_1^{(0)}, f_2^{(0)}, \ldots \}$ is given. Suppose also that a collection $\mathfrak{F}_i = \{ f_1^{(i)}, f_2^{(i)}, \ldots \}$, $i > 0$, can be obtained as follows: fix $f_{k_{i-1}}^{(i-1)} \in \mathfrak{F}_{i-1}$, $k_i \neq k_j$ for $i \neq j$, and multiply the words $f_l^{(i-1)}$ by words $t_l^{(i-1)}$ (possibly equal to identity words), each of which is a consequence of the word $f_{k_{i-1}}^{(i-1)}$, where $t_l^{(i-1)} = 1$ for $l \leqslant k_{i-1}$; finally, suppose $f_l^{(i)} = t_l^{(i-1)} f_l^{(i-1)}$.*

Then each sequence $f_s^{(0)}, f_s^{(1)}, f_s^{(2)}, \ldots$ stabilizes at some $f_s^{(n_s)} = f_s$, and the collection $\mathfrak{F} = \{ f_1, f_2, \ldots \}$ is equivalent to $\mathfrak{F}_0$.

PROOF. It is clear from the definition of $\mathfrak{F}_i$ that $f_s^{(i)} \neq f_s^{(i-1)}$ only if $k_{i-1} < s$. Since $k_i \neq k_j$ for $i \neq j$, the sequence $f_s^{(0)}, f_s^{(1)}, \ldots$ contains at most $s - 1$ pairs of distinct adjacent terms, i.e. it stabilizes. To prove the second assertion, note that the reverse transitions $\mathfrak{F}_i \to \mathfrak{F}_{i-1}$ can be effected by the same transformations as $\mathfrak{F}_{i-1} \to \mathfrak{F}_i$, and the finite collections $\{ f_1^{(i-1)}, \ldots, f_r^{(i-1)} \}$ and $\{ f_1^{(i)}, \ldots, f_r^{(i)} \}$ are equivalent for any r.

We now use the fact that for any r there exists a number n such that $f_1^{(n)} = f_1, \ldots, f_r^{(n)} = f_r$, and the equivalence of $\mathfrak{F}$ and $\mathfrak{F}_0$ then follows from the equivalence of $\{ f_1, \ldots, f_r \}$ and $\{ f_1^{(0)}, \ldots, f_r^{(0)} \}$ for an arbitrary r.

Theorem 1 is obvious if $\mathfrak{W}$ is finitely based. We therefore assume that the basis of identities

$$f_1^{(0)}(x_1, \ldots, x_{n_1}) \equiv 1, \qquad f_2^{(0)}(x_1, \ldots, x_{n_2}) \equiv 1, \ldots \qquad (2)$$

of the variety $\mathfrak{W}$ is not equivalent to a finite basis. Suppose $\mathfrak{U}$ is the variety in Lemma 1, and V and U are the verbal subgroups of F corresponding to these varieties. By a simultaneous induction we define:

a) a descending series of verbal subgroups $V = V_0 \supset V_1 \supset V_2 \supset \cdots \supset U$ such that V_i is maximal in V_{i-1};

b) an ascending series of verbal subgroups $U = U_0 \subset U_1 \subset U_2 \subset \cdots \subset V$, corresponding to certain finitely based varieties, such that

$$V_i \supset V_{i-1} \cap U_{i-1}, \qquad i = 1, 2, \ldots; \tag{3}$$

c) finite subsets $\mathfrak{N}_i$ of positive integers, where $\mathfrak{N}_0 = \varnothing$;

d) systems of identities $\mathfrak{F}_i = \{ f_1^{(i)}, f_2^{(i)}, \ldots \}$ such that $f_k^{(i)} \in V_i$ for $k \notin \mathfrak{N}_i$; and

e) words $u_i \in V_{i-1} \backslash V_i$, $i = 1, 2, \ldots$.

Suppose $i > 0$. We carry out the inductive step.

1) Note that $V_{i-1} \neq V_{i-1} \cap U_{i-1}$, as can be seen from Lemma 3. Therefore, by Lemma 2, there exists a maximal verbal subgroup V_i of V_{i-1} such that (3) holds. Obviously, $V_i \supset U$.

2) In the system $\mathfrak{K}_i = \{ f_{d_1}^{(i-1)}, f_{d_2}^{(i-1)}, \ldots \}$ obtained from $\mathfrak{F}_{i-1}$ by deleting all words $f_s^{(i-1)}$, where $s \in \mathfrak{N}_{i-1}$, we choose the word $f_{k_{i-1}}^{(i-1)}$ with the smallest index that does not lie in V_i, and put $\mathfrak{N}_i = \mathfrak{N}_{i-1} \cup \{ k_{i-1} \}$. If there is no such word, we put $\mathfrak{N}_i = \mathfrak{N}_{i-1}$.

3) If $\mathfrak{N}_i = \mathfrak{N}_{i-1}$, we put $\mathfrak{F}_i = \mathfrak{F}_{i-1}$. Otherwise, fix the word $f_{k_{i-1}}^{(i-1)}$ defined above. By the inductive assumption, all words $f_k^{(i-1)}$ in $\mathfrak{K}_i$ lie in V_{i-1}. If $l < k_{i-1}$, $l \notin \mathfrak{N}_{i-1}$, then, by choice of k_{i-1}, $f_l^{(i-1)} \in V_i$. Note that V_{i-1} modulo V_i is generated (as a verbal subgroup) by the word $f_{k_{i-1}}^{(i-1)}$, in view of the maximality of V_i in V_{i-1}. Therefore, if $l > k_{i-1}$, $l \notin \mathfrak{N}_{i-1}$, we have $f_l^{(i-1)} = (t_l^{(i-1)})^{-1} f_l^{(i)}$, where the $t_l^{(i-1)}$ are consequences of the word $f_{k_{i-1}}^{(i-1)}$, and $f_l^{(i)} \in V_i$. Putting $f_l^{(i)} = f_l^{(i-1)}$ for $l \in \mathfrak{N}_i$ or $l \leqslant k_{i-1}$, and $f_l^{(i)} = t_l^{(i-1)} f_l^{(i-1)}$ for $l > k_{i-1}$, $l \notin \mathfrak{N}_i$, we have $f_k^{(i)} \in V_i$ for $k \notin \mathfrak{N}_i$.

4) If $\mathfrak{N}_i \neq \mathfrak{N}_{i-1}$, we put $u_i = f_{k_{i-1}}^{(i-1)}$; otherwise we choose u_i arbitrarily from the set $V_{i-1} \backslash V_i$.

5) We define the subgroup U_i to be the smallest verbal subgroup containing U_{i-1} and u_i. It is clear that U_i also admits a finite basis (as a verbal subgroup).

Having completed this construction, we define word w_i as follows. The transition from $\mathfrak{F}_{i-1}$ to $\mathfrak{F}_i$ was carried out so as to admit the hypothesis of Lemma 4. In that case, let $\mathfrak{F} = \{ f_1, f_2, \ldots \}$ be the system whose existence follows from that lemma. Consider the subsystem (if it is finite, augment it by an infinite number of identity words) $\mathfrak{F}' = \{ f_{j_1}, f_{j_2}, \ldots \}$ consisting of all words f_{j_r} such that $j_r \in \mathfrak{N} = \bigcup_0^\infty \mathfrak{N}_d$, and rewrite each word of this system in a new set of variables $Y = \{ y_1, y_2, \ldots \}$. Finally, put

$$w_i = u_i(X) f_{j_i}(Y). \tag{4}$$

LEMMA 5. *The variety $\mathfrak{W}'$ defined by the system $\{ w_1, w_2, \ldots \}$ is equal to $\mathfrak{W}$.*

PROOF. It follows from the definition of the words u_i and f_k that each $w_i \in V$, i.e. $\mathfrak{W} \supseteq \mathfrak{W}'$. On the other hand, for $k_{i-1} \in \mathfrak{N}_i$ we have $f_{k_{i-1}}^{(i-1)} = f_{k_{i-1}}$, since under the transition $\mathfrak{F}_{i-1} \to \mathfrak{F}_i$ (and subsequent ones) the words with indices in $\mathfrak{N}_i$ do not change. Therefore, some word in $\mathfrak{F}$ either is equal to $u_i = f_{k_{i-1}}^{(i-1)}$ and is thus a consequence of w_i (u_i and f_{j_i} depends on disjoint sets of variables), or else is equal to some f_{j_i} in $\mathfrak{F}'$ and is also a consequence of some w_i. The reverse inclusion is proved: $\mathfrak{W} \subseteq \mathfrak{W}'$.

LEMMA 6. *The system of words* (4) *is independent.*

PROOF. As we have already observed, if $s \in \mathfrak{N}_d \backslash \mathfrak{N}_{d-1}$, then $f_s = f_s^{(d-1)}$. If $s \notin \mathfrak{N}$, then the sequence $f_s^{(0)}, f_s^{(1)}, f_s^{(2)}, \ldots$, as is easily seen, can stabilize only if, for some l, $f_s^{(l)} \in \bigcap_0^\infty V_i$ (and $f_s = f_s^{(l)}$). Therefore, it follows from (4) and the definition of the words u_i that

$$w_i \in V_{i-1} \backslash V_i. \tag{5}$$

Assume that w_i follows from $w_1, \ldots, w_{i-1}, w_{i+1}, \ldots$. Then there exists a representation $w_i = ab$, where a is a consequence of the words $w_1, \ldots, w_{i-1}$ and b is a consequence of the words $w_{i+1}, w_{i+2}, \ldots$. Equality (4) and the definition of the subgroup U_{i-1} show that $a \in U_{i-1}(\bigcap_0^\infty V_k) \subseteq U_{i-1}V_i$. It follows from (5) that $b \in V_i$. Also, from (5) we obtain $a = w_i b^{-1} \in V_{i-1}V_i = V_{i-1}$, i.e $a \in U_{i-1}V_i \cap V_{i-1} = $ (by the modular law) $(U_{i-1} \cap V_{i-1})V_i = $ (see (3)) V_i, i.e. $w_i = ab \in V_i V_i = V_i$, contrary to (5). This contradiction proves Lemma 6 and, taken together with Lemma 5, also Theorem 1.

REMARK 2. The solvability of the variety $\mathfrak{W}$ was used only in Lemma 1. In general, in the presence of an infinite descending chain

$$\mathfrak{W}_1 \supset \mathfrak{W}_2 \supset \cdots; \quad \bigcap_{i=1}^\infty \mathfrak{W}_i = \mathfrak{W} \tag{6}$$

such that the varieties $\mathrm{var}(F_i(\mathfrak{W}_1))$ (where $F_i(\mathfrak{W}_1)$ is a free group of rank i in the variety $\mathfrak{W}_1$) satisfy the descending chain condition for subvarieties, we can prove, as above, that the variety $\mathfrak{W}$ is distinguished within $\mathfrak{W}_1$ (and therefore in the class of all groups, as can easily be shown) by an independent system of identities.([2]) In particular, this implies the result of Kovács [8], who proved under the same assumption on $\mathfrak{W}_1$ that there are continuum many varieties included between $\mathfrak{W}_1$ and $\mathfrak{W}$.

The varieties $\mathrm{var}(F_i(\mathfrak{W}_1))$ are Cross if $\mathfrak{W}_1$ is locally finite, and hence $\mathfrak{W}_1$ satisfies the Kovács condition. In that case, an infinite chain (6) obviously exists if $\mathfrak{W}_1$ has a finite basis of identities (and system (2) is not equivalent to a finite system).

It is true that each locally finite variety is contained in some finitely based, locally finite variety? This is a well-known open question. It was answered affirmatively, for example, for varieties of A-groups [9], to which, therefore, Theorem 1 extends. We have actually proved that each locally finite variety can be defined in the class of locally finite groups by an independent system of identities. An analogous remark holds for locally nilpotent varieties. (It is necessary to invoke Lyndon's theorem on a finite basis of identities of nilpotent varieties [1], 34.14.)

([2]) In this formulation Theorem 1 can also be proved for arbitrary varieties of Ω-groups. An example of a variety of semigroups that cannot be defined by any independent system of identities was recently contructed by A. N. Trahtman.

REMARK 3. Using the analog of Lyndon's theorem for nilpotent varieties of algebras and the theorem of Mal'cev [10] that relatively free algebras over an infinite field can be approximated by nilpotent algebras, we can prove without difficulty that an arbitrary variety of (not necessarily associative) algebras over an infinite field can be defined by an independent system of identities.

The analog of Theorem 1 for locally finite varieties of Lie rings can be proved by using a result of [11].

A verbal subgroup V of F that is not finitely based can be represented as the union $V = \bigcup_1^\infty V_i$ of the terms of an infinite ascending chain of verbal subgroups

$$V_1 \subset V_2 \subset \cdots \subset V_n \subset \cdots . \tag{7}$$

We will show how to associate with each such chain a certain infinite irredundant system of verbal subgroups (it was constructed by the author back in 1968 [12]). As always, $U(G)$ denotes the verbal subgroup of G corresponding to the variety $\mathfrak{U}$; the verbal subgroup corresponding to the variety $\mathfrak{A}_n$ of Abelian groups of exponent n is denoted by $A_n(G)$. The commutator subgroup of G is denoted by G'.

THEOREM 2. *For an arbitrary fixed prime p there exists an infinite sequence of natural numbers $k_1 = k_1(p), k_2 = k_2(p), \ldots$ such that the system $\{A_{p^i}(V_{k_i})\}_{i=1,2,\ldots}$ is irredundant (the V_i are taken from the chain (7)).*

LEMMA 7. *Suppose we are given a group G and an ascending chain of subgroups $H_1 \subseteq H_2 \subseteq \cdots$. Then $\bigcup_i U(H_i) = U(\bigcup H_i)$.*

The proof is obvious.

LEMMA 8. *If G is a free group in the variety $\mathfrak{A}_{p^n}$, $n \geq 1$, and p a prime, then the lower layer of G is equal to $A_{p^{n-1}}(G)$.*

PROOF. The assertion is true because G is a direct product of cyclic groups of order p^n.

LEMMA 9. *Suppose F_∞ is a free group of infinite rank in some variety and $V = V(F_\infty)$ is a nonidentity verbal subgroup. Then V is infinite.*

PROOF. Suppose $\{x_\alpha\}_{\alpha \in M}$ is a system of free generators of F_∞. If $1 \neq v_1 = v(x_{\alpha_1}, \ldots, x_{\alpha_n}) \in V$, then $V \ni v_2 = v(x_{\beta_1}, \ldots, x_{\beta_n})$, and so it is possible to obtain infinitely many distinct elements $v_i \in V$, writing them in terms of disjoint sets of free variables.

LEMMA 10 (ŠMEL'KIN [13]). *Suppose F is a free group, R a normal subgroup of infinite index, and $V(R)$ a verbal subgroup of R, where $V(R) \neq R$. Then the factor group $F/V(R)$ has trivial center.*

LEMMA 11. *For an arbitrary fixed prime p there exists an infinite sequence of natural numbers $k_1 = k_1(p)$, $k_2 = k_2(p), \ldots$ such that the following conditions are satisfied*:

1) $A_p(V_{k_1}) \not\subseteq A_{p^2}(V)$,
2) $A_p(V_{k_1}) \cap A_{p^2}(V_{k_2}) \not\subseteq A_{p^3}(V)$,

$\cdots$

i) $\bigcap_{s=1}^{i} A_{p^s}(V_{k_s}) \subseteq A_{p^{i+1}}(V)$.

$\cdots$

PROOF. Since $A_p(V) \not\subseteq A_{p^2}(V)$, it follows from Lemma 7 that there exists a number k_1 such that condition 1) holds.

Suppose numbers $k_1 < k_2 < \cdots k_{i-1}$ have already been chosen so that conditions 1), 2),$\ldots,i-1$) hold. Put $B = \bigcap_{s=1}^{i-1} A_{p^s}(V_s)$. According to condition $i-1$), $B \not\subseteq A_{p^i}(V)$, and a fortiori $B \not\subseteq A_{p^{i+1}}(V)$; and since $A_{p^i}(B) \subseteq A_{p^{i+1}}(V)$, there exists an element $b \in B$ such that $b \notin A_{p^{i+1}}(V)$, but $b^p \in A_{p^{i+1}}(V)$. Since $b \in B \subset V$, it follows from Lemma 8 that $b \in A_{p^i}(V)$. Finally, by Lemma 7, there exists a natural number $k_i > k_{i-1}$ such that $B \cap A_{p^i}(V_{k_i}) \ni b$; hence condition i) holds. Thus, Lemma 11 is proved by induction.

Now, having chosen the numbers k_i, we put $U_i = A_{p^i}(V_{k_i})$, $i = 1, 2, \ldots$, and $U_i^+ = \bigcap_{j>i} U_j$ and $U_i^- = \bigcap_{j<i} U_j$; we will show that the system $\{U_i\}_{i=1,2,\ldots}$ is irredundant, i.e.

$$U_i^+ \cap U_i^- \not\subseteq U_i. \tag{8}$$

Suppose first that $i > 1$. Since $U_i \subseteq A_{p^i}(V)$, it follows from condition $i-1$) of Lemma 11 that

$$U_i^- \not\subseteq U_i. \tag{9}$$

We have the inclusions

$$U_i^+ = \bigcap_{j>i} U_j = \bigcap_{j>i} A_{p^j}(V_{k_j}) \supseteq \bigcap_{j>i} V_{k_j}' = V_{k_{i+1}}' \supseteq [U_i^-, V_{k_{i+1}}]. \tag{10}$$

Moreover, it is obvious that $U_i^- \supseteq [U_i^-, V_{k_{i+1}}]$; hence $U_i^- \cap U_i^+ \supseteq [U_i^-, V_{k_{i+1}}]$. If (8) were false, we would have $U_i \supseteq [U_i^-, V_{k_{i+1}}]$. This would mean that the factor group $V_{k_{i+1}}/U_i$ has nontrivial center, in view of (9). However, this contradicts Lemma 10, since $V_{k_{i+1}}/U_i = V_{k_{i+1}}/A_{p^i}(V_{k_i})$ and, by Lemma 9, $V_{k_{i+1}}/V_{k_i}$ is infinite.

Relation (8) is proved for $i > 1$. It remains to observe that $\bigcap_{i>1} U_i \supseteq \bigcap_{i \geqslant 2} V_{k_i}' = V_{k_2}'$, and if $\bigcap_{i>1} U_i \subseteq U_1$, then the factor group V_{k_2}/U_1 is Abelian, which again is refuted by Lemmas 9 and 10. Theorem 2 is proved.

An analog of Theorem 2 also holds for varieties of Lie algebras over a field of characteristic 0.

The groups constructed in [2] and satisfying condition (1) are "hypocritical" in the sense of [14], as can be seen from Theorem 9.1 of [14]. It follows that var(G_i, $i = 1, 2, \ldots$) cannot be decomposed into a finite join of indecomposable varieties. In conclusion, we observe that the problem of Tarski mentioned at the beginning of this paper has something in common with the following question: Can each variety be decomposed into a (finite or infinite) irredundant join of indecomposable varieties?

Received 25/SEPT/75

BIBLIOGRAPHY

1. Hanna Neumann, *Varieties of groups*, Springer-Verlag, 1967.

2. A. Ju. Ol'šanskiĭ, *On the problem of a finite basis of identities in groups*, Izv. Akad. Nauk SSSR Ser. Mat. **34** (1970), 376–384; English transl. in Math. USSR Izv. **4** (1970).

3. S. I. Adjan, *Infinite irreducible systems of group identities*, Izv. Akad. Nauk SSSR Ser. Mat. **34** (1970), 715–734; English transl. in Math. USSR Izv. **4** (1970).

4. *The Kourovka notebook: Unsolved problems in group theory*, 4th ed., Inst. Mat. Sibirsk. Otdel. Akad. Nauk SSSR, Novosibirsk, 1973. (Russian)

5. M. R. Vaughan-Lee, *Uncountably many varieties of groups*, Bull. London Math. Soc. **2** (1970), 280–286.

6. Graham Higman, *The orders of relatively free groups*, Proc. Internat. Conf. Theory of Groups (Canberra, 1965), Gordon and Breach, New York, 1967, pp. 153–165.

7. A. Ju. Ol'šanskiĭ, *On the orders of free groups of locally finite varieties*, Izv. Akad. Nauk SSSR Ser. Mat. **37** (1973), 89–94; English transl. in Math. USSR Izv. **7** (1973).

8. L. G. Kovács, *On the number of varieties of groups*, J. Austral. Math. Soc. **8** (1968), 444–446.

9. Roger M. Bryant, *On the laws of the variety $s\mathfrak{A}_e$*, J. Austral. Math. Soc. **14** (1972), 364–367.

10. A. I. Mal'cev, *On algebras defined by identities*, Mat. Sb. **26** (**68**) (1950), 19–33. (Russian)

11. Ju. A. Bahturin and A. Ju. Ol'šanskiĭ, *Identical relations in finite Lie rings*, Mat. Sb. **96** (**138**) (1975), 543–559; English transl. in Math. USSR Sb. **25** (1975).

12. Tenth All-Union Algebraic Colloq. (Novosibirsk, 1969), Abstracts of Reports and Papers, Vol. 1, Inst. Mat. Sibirsk. Otdel. Akad. Nauk SSSR, Novosibirsk, 1969. (Russian)

13. A. L. Šmel'kin, *Wreath products and varieties of groups*, Izv. Akad. Nauk SSSR Ser. Mat. **29** (1965), 149–170. (Russian)

14. L. F. Harris, *Varieties and section closed classes of groups*, Ph.D. Thesis, Australian National Univ., Canberra, 1973.

Translated by G. A. KANDALL

Amer. Math. Soc. Transl.
(2) Vol. **119**, 1983

Radicals in Groups, Operations
on Classes of Groups,
and Radical Classes*

B. I. PLOTKIN

In memory of Anatonliĭ Ivanovič Mal'cev

Introduction

The idea of a radical is one of the most fruitful organizing principles in the theory of groups. Two fundamental aspects can be distinguished in the theory of radicals. On the one hand, radicals in groups are important characteristic subgroups, which play a substantial role in the structure theory of individual groups. In particular, the proofs of many group-theoretical theorems are simplified considerably when radicals are brought in. This is the applied role of radicals. However, it is known that the study of individual groups is closely interwoven with a discussion of classes of groups, and in recent years the outlines of a new large branch of group theory have become visible: the general theory of classes of groups. Various operations on classes can be defined, which enable us to talk in this language of the structure of classes of groups, and various systems of classes can be analyzed that are well organized by certain operations on classes.([1]) In the theory of classes a prominent place is occupied by radical classes, and here lies the second aspect of the theory of radicals.

In this paper we place new accents on the concepts of a radical and a radical class. Throughout we regard a radical as a group-theoretical function with certain special properties. A radical in an individual group is then the value of this function. Between radicals as functions and radical classes of groups there is a clearly defined one-to-one correspondence. In the theory of radical classes the

---

1980 *Mathematics Subject Classification.* Primary 20E10, 20E34; Secondary 06A15.

* Translation of Selected Questions of Algebra and Logic (A. I. Mal'cev Memorial Coll.), "Nauka", Novosibirsk, 1973, pp. 205–244; MR **52** #10901.

([1]) The outstanding contribution of A. I. Mal'cev to the general theory of classes of algebraic systems is common knowledge.

language of operators on classes of groups is used systematically, and the operators of the radical closure of classes are investigated. Certain algebraic operations in the system of all radical classes are being studied, and here the outlook is close, for example, to the analogous parts of the theory of varieties of groups. The system of all varieties of groups is known to be a set, and this set is a free semigroup under multiplication of classes. The system of all radical classes is not a set, but it is also closed under multiplication, and here multiplication is associative. Even infinite products of radical classes can be considered. Simultaneously with radicals we also study coradicals, and in what follows we point out some new phenomena of duality between radicals and coradicals. Some other special features are also mentioned.

Operations on classes lead to a substantial enrichment of the available radicals and coradicals, which are useful in applications. In concrete situations we can choose whatever radical or coradical suits our taste and fits the prevailing conditions.

The first papers on the general theory of radicals, due to Kuroš [1] and Amitsur [2], were published at the beginning of the fifties. Kuroš placed the main emphasis on the radical class. Although both these papers were written largely under the influence of the ring-theoretical situation, their value was considerably wider: radicals began to penetrate actively into other classes of algebraic systems. Particular attention was paid to radicals in groups. In subsequent publications Kuroš and other authors made it clear that the role of radicals in groups is in many respects different from their role, say, in rings. Furthermore, in the theory of groups the idea of a radical began to compete with various other organizing principles. All this manifests itself in the axiomatics of radicals in groups; but apparently there is not yet any stable axiom system. It seems to us that in this paper we have chosen a certain optimal version. In particular, we shall see that it is advantageous to split the concept of a radical into that of a preradical, radical, and strict radical. We mention, further, that the inverse process also takes place: the development of the theory of radicals in groups influences the general theory of radicals in algebraic systems. Some constructions of this paper can likewise be transferred appropriately. Occasionally this is a trivial matter, but in a number of cases the scheme changes substantially.

The paper is concerned, on the whole, with a clarification of the situation; it can also be regarded as a survey of some branches of the theory of radicals in groups, a survey with new positions taken into account. It does not contain information on the whole theory of radicals in groups, but a fairly complete outline can be obtained by considering it in conjunction with the survey on radicals in the Appendix in Kuroš's book [17], and also the corresponding sections (§§5.2–5.4) in the author's book [4]. Proofs of results that are known or can easily be derived from known results are, as a rule, omitted. Some results from the author's papers [5] and [6] are also quoted without proofs. Finally, a number of problems are stated in the paper.

§1. Operations on classes of groups, and closure operators

1. *Operators.* Many important operations on classes of groups have been accumulated over the years, and in a number of cases there are interesting relations between these operations, which lead to a certain calculus of operations. Here we have in mind, in the first instance, operators that associate with one class of groups certain other classes. The study of relations between such operators in its pure form is essentially an integral part of the theory of groups, indeed, one of its organizing principles. In explicit form this fact was emphasized recently by P. Hall [7], and has by now found widespread recognition. The language of operations on classes of groups is equally convenient in the theory of radicals.

Let us explain a few initial concepts. The main source is Hall's paper mentioned above, but there are also some concepts that have not been noted before explicitly.

By $\mathfrak{X}$, $\mathfrak{X}_1,\ldots$ and other convenient letters we denote variables ranging over classes of groups. For various specific classes, which occupy a special position in the theory of groups, we use a standard system of special notation. For example, $\mathfrak{A}$ is the class of all Abelian groups. All classes to be considered are abstract; that is, together with any group the class also contains all groups isomorphic to it. We assume, furthermore, that all the classes contain a trivial group.

It is natural to speak of the intersection and union of classes. Apart from these structural operations we can also consider various special operations, which are due to the specific group-theoretical character. These can be operations (functions of several variables), or unary operations (operators). For some of these operations, which play a particularly prominent role, we can use suitable special symbols. Variable operators are denoted by U, V, $\mathsf{U}_1,\ldots$, and the result of applying them to a class $\mathfrak{X}$ by $\mathsf{U}\mathfrak{X}$ or, occasionally, $\mathsf{U}[\mathfrak{X}]$. Composition (superposition) of operations is defined in the usual way. Apart from finite compositions of operations we can here also consider infinite compositions in which unions or intersections of classes take part. Suppose, for example, that U is an operator with the property of *class extension*. By this we mean that the following two conditions are satisfied: $\mathfrak{X} \subset \mathsf{U}\mathfrak{X}$ for any $\mathfrak{X}$, and if $\mathfrak{X}_1 \subset \mathfrak{X}_2$, then $\mathsf{U}\mathfrak{X}_1 \subset \mathsf{U}\mathfrak{X}_2$. The powers of such a U for ordinal numbers α are defined as follows: $\mathsf{U}^0\mathfrak{X} = \mathfrak{X}$, $\mathsf{U}^1 = \mathsf{U}$ and $\mathsf{U}^{\alpha+1}\mathfrak{X} = \mathsf{U}(\mathsf{U}^\alpha\mathfrak{X})$; and if α is a limit number, then $\mathsf{U}^\alpha\mathfrak{X}$ is the union of the classes $\mathsf{U}^\beta\mathfrak{X}$ with $\beta < \alpha$. We now define an operator $\overline{\mathsf{U}}$ by taking $\overline{\mathsf{U}}\mathfrak{X}$ to be the union of all $\mathsf{U}^\alpha\mathfrak{X}$ for all transfinite ordinals α. It is clear that all powers of U and $\overline{\mathsf{U}}$ are extension operators. We write $\mathsf{U} < \mathsf{V}$ if $\mathsf{U}\mathfrak{X} \subset \mathsf{V}\mathfrak{X}$ for any $\mathfrak{X}$. Evidently, if U is an extension operator, then $\alpha < \beta$ implies that $\mathsf{U}^\alpha < \mathsf{U}^\beta$.

A class $\mathfrak{X}$ is said to be *closed* under an extension operator U if $\mathsf{U}\mathfrak{X} = \mathfrak{X}$. On the other hand, an extension operator U is a *closure operator* if every class $\mathsf{U}\mathfrak{X}$ is U-closed. This is equivalent to the relation $\mathsf{U}^2 = \mathsf{U} = \overline{\mathsf{U}}$. An operator U is called *bounded* if $\mathsf{U}^{\alpha+1} = \mathsf{U}^\alpha = \overline{\mathsf{U}}$ for some α. The minimal α with this property is called the *order* (*bound*) of U.

As a rule, all the extension operators we have to consider satisfy the following *localization condition*:

If $\mathfrak{X}$ is a class of groups and $G \in U\mathfrak{X}$, then $\mathfrak{X}$ contains a subclass $\mathfrak{X}'$ which is the abstract closure of a certain set of groups and for which $G \in U\mathfrak{X}'$.

We assume that there is a method of distinguishing sets and classes; in particular, we assume that every set has a cardinality. The abstract closure of a set of groups (that is, the class of all groups isomorphic to the groups of the given set) is also briefly referred to as a set of groups. In what follows it will be necessary, not withstanding existing prohibitions, to combine even classes in some system of higher order: superclasses. All such "invalid" operations are easy to explain in specific situations. On the other hand, apparently a justification can also be given in terms of a suitable axiomatic.

Now we come to our first proposition.

PROPOSITION 1. *If an operator U satisfies the localization condition, then for every $\mathfrak{X}$ the class of groups $\overline{U}\mathfrak{X}$ is the minimal U-closed class containing $\mathfrak{X}$.*

PROOF. First of all, we observe that for every extension operator U the intersection of the U-closed classes is U-closed; therefore we can speak of the minimal U-closed class containing a given class $\mathfrak{X}$. If U is a closure operator, then this minimal class is $U\mathfrak{X}$. In general, it is clear that if $\mathfrak{X}_1$ is a U-closed class containing $\mathfrak{X}$, then, since the operator is monotonic, this class also contains all $U^\alpha\mathfrak{X}$, and therefore $\overline{U}\mathfrak{X}$.

To prove the proposition it remains to verify that under our condition $\overline{U}\mathfrak{X}$ is a U-closed class. Suppose that a group G belongs to $U(\overline{U}\mathfrak{X})$, and let $\mathfrak{X}'$ be a set of groups belonging to $\overline{U}\mathfrak{X}$ and such that $G \in U\mathfrak{X}'$. It is clear that for some α all groups in $\mathfrak{X}'$ belong to $U^\alpha\mathfrak{X}$. Here the class $U\mathfrak{X}'$ is contained in $U^{\alpha+1}\mathfrak{X}$. Consequently, $G \in \overline{U}\mathfrak{X}$, as required.

We have seen that $\overline{U}$ is a closure operator; in fact, the minimal closure operator containing U. For this reason $\overline{U}$ is called the *closure* of U. It is not hard to observe that the localization condition is essential: without it $\overline{U}$ need not be a closure operator.

Apart from extension operators it is sometimes useful to consider another type of monotonic operators, the *restriction operators*; that is, operators U such that $U\mathfrak{X} \subset \mathfrak{X}$ (and $\mathfrak{X}_1 \subset \mathfrak{X}_2$ implies that $U\mathfrak{X}_1 \subset U\mathfrak{X}_2$). To define the powers of such restriction operators it is natural to start out from the intersection of classes. But we are mainly interested in extension operators.

By analogy with the closure of a single extension operator we can also speak of the closure of a system of such operators. For example, if U and V are two extension operators, then their joint closure is the operator $\{U, V\}$ that associates with $\mathfrak{X}$ the class $\{U, V\}\mathfrak{X}$, which among the classes containing the original $\mathfrak{X}$ is the minimal U- and V-closed class. $\{U, V\}$ is a closure operator. Similarly we can speak of the operator $\{U_1, U_2, \ldots\}$, and the number of terms can here also be infinite. For a single generator we have $\{U\} = \overline{U}$, provided that U satisfies the localization condition.

One of the important questions arising is the structure of the operator $\{U_1, U_2, \ldots\}$ in dependence on the constituents $U_1, U_2, \ldots$. We give here a rough rule of computation. Let U_α, $\alpha \in I$, be a collection of extension operators, and suppose they all satisfy the localization condition. We denote by U their compositum: $U\mathfrak{X}$ is the union of the classes $U_\alpha\mathfrak{X}$. It is not hard to see that $\{\ldots, U_\alpha, \ldots\} = \overline{U}$. If the original operators are finite in number, then instead of their compositum we can take for U also the product of all these operators in an arbitrary order and obtain the analogous rule.

Naturally, various simplifications result from the presence of additional relations between operators. In this context we mention the following simple and useful facts [7]. Let U and V be closure operators. Then the equality $\{U, V\} = UV$ is equivalent to the relation $VU \leqslant UV$. Furthermore, the following three conditions are equivalent: 1) $V \leqslant U$, 2) $U = \{U, V\}$, and 3) every U-closed class is also V-closed.

We shall apply these rules repeatedly.

Many interesting types of classes of groups can be defined by the property of being closed under certain specific collections of operators on classes. This refers, for example, to varieties of groups, and to radical and coradical classes. We now define the relevant operators.

By $H\mathfrak{X}$ (Hom $\mathfrak{X}$) (^{2}) we denote the class of all groups that are homomorphic images of groups in $\mathfrak{X}$. Also, $S\mathfrak{X}$ (Sub $\mathfrak{X}$) denotes the class of all groups that can be embedded as subgroups in some groups of $\mathfrak{X}$, $C\mathfrak{X}$ (Cart $\mathfrak{X}$) is the class of all Cartesian (complete direct) products of groups in $\mathfrak{X}$, and $R\mathfrak{X}$ is the class of groups that are generated by their normal $\mathfrak{X}$-subgroups. The operator CoR works dually: $\mathsf{CoR}\,\mathfrak{X}$ is the class of groups that are residually $\mathfrak{X}$-groups. In other words, $G \in \mathsf{CoR}\,\mathfrak{X}$ if G has a system of normal subgroups A_α such that always $G/A_\alpha \in \mathfrak{X}$ and the intersection of all A_α is trivial. By $I\mathfrak{X}$ we denote the class of all groups that can be embedded as normal subgroups in a group of the class $\mathfrak{X}$.

Evidently, H, S, C, and CoR are closure operators, but R and I are not. It is easy to see that $\bar{I} = I^\omega$ and that $\bar{I}\mathfrak{X}$ is the class of groups that are embeddable as subnormal subgroups in a group of the class $\mathfrak{X}$. Later we shall see that $\bar{R} = R^{\omega+1}$ and that $\bar{R}\mathfrak{X}$ is the class of groups that can be generated by their subnormal $\mathfrak{X}$-subgroups.

It is well known that a class of groups is a variety if and only if it is closed under the operators H, S, and C. We write $V = \{H, S, C\}$, and then for every $\mathfrak{X}$ the class $V\mathfrak{X}$ is the variety generated by $\mathfrak{X}$. In this context it is desirable to calculate the explicit form of the operator V. It is easy to see that the following relations hold: $SH < HS$, $CS < SC$, and $CH < HC$. By means of the rule mentioned above we immediately deduce that $\{H, S, C\} = HSC$. Another expression for V can be obtained by means of a recent result of Kogalovskiĭ [8], according to which $\{H, S, C\} = \{H, \mathsf{CoR}\}$. It is equally easy to see that

---

(^{2}) This used to be written as $\mathbf{Q}\mathfrak{X}$.

CoRH < HCoR, and therefore V = HCoR. So we arrive at a curious relationship between operators: HSC = HCoR.([3])

We say that a class of groups $\mathcal{X}$ is a *radical class* if it is R- and H-closed. We write Rad = {R, H}. It is fairly obvious that H$\overline{\text{R}}$ < $\overline{\text{R}}$H; hence Rad = $\overline{\text{R}}$H. Thus, if $\mathcal{X}$ is a certain class, then the radical class Rad $\mathcal{X}$ is $\overline{\text{R}}$H$\mathcal{X}$. A radical class $\mathcal{X}$ is called *hereditarily* radical if I$\mathcal{X}$ = $\mathcal{X}$. It would, therefore, be desirable to compute the operator {R, H, I}. However, this is by no means a simple matter.

A *coradical class* is a class of groups that is closed under the operators CoR and I (= CoH). From $\overline{\text{I}}$CoR < CoR$\overline{\text{I}}$ we deduce that {CoR, I} = CoR$\overline{\text{I}}$. Obviously, if $\mathcal{X}$ is a coradical class, then H$\mathcal{X}$ is a variety.

Turning to the definition of varieties, we recall that the operator V has the following convenient characterization: V$\mathcal{X}$ is the class of all groups on which all the identities of the class $\mathcal{X}$ hold. When we consider instead of identities certain other types of formulas, say, in the language of the lower predicate calculus (LPC), we can specify in a similar manner also certain other operators (see, for example, [32]). Next, we recall the following important feature of varieties. Every variety can be generated by a single group, a suitable free group.

The other approach leads to certain new problems. For example, let U be an operator. We say that two classes of groups $\mathcal{X}_1$ and $\mathcal{X}_2$ are U-*equivalent* if U$\mathcal{X}_1$ = U$\mathcal{X}_2$. We can also speak of U-equivalence of individual groups. A number of interesting problems are connected with the investigation of radical equivalence of groups.

If U is an operator and $\mathcal{X}$ a class of groups, it is natural to note the following characteristic feature of $\mathcal{X}$ relative to U: the minimal α (if it exists) with the property U$^\alpha\mathcal{X}$ = U$^{\alpha+1}\mathcal{X}$. If $\mathcal{X}$ is U-closed, then the corresponding α is zero, and in the general case α measures the deviation from being U-closed. Next, let G be a group belonging to the class $\overline{\text{U}}\mathcal{X}$. We may associate with G the smallest ordinal number α for which $G \in \text{U}^\alpha\mathcal{X}$. We call this α the U-*step* of G over $\mathcal{X}$. Of course, this step cannot be a limit number. The study of such steps in concrete situations often turns out to be an interesting matter.

2. *Multiplication of classes.* An important example of a binary operation on classes of groups is the multiplication: if $\mathcal{X}_1$ and $\mathcal{X}_2$ are two classes, their product $\mathcal{X}_1\mathcal{X}_2$ is the class consisting of groups that are extensions of groups in $\mathcal{X}_1$ by groups in $\mathcal{X}_2$. A generalization of this operation is the n-argument operation that associates with an ordered collection of classes $\mathcal{X}_1,\ldots,\mathcal{X}_n$ the class $\mathcal{X}_1\ldots\mathcal{X}_n$, which consists of the groups having a normal series of length n in which the ith factor, $1 \leqslant i \leqslant n$, is a group in $\mathcal{X}_i$. Similarly we can define an n-argument operation on the basis of invariant series. If all the classes $\mathcal{X}_i$ are taken as coincident, we arrive at certain special one-argument operations.

It is appropriate to consider also infinite-argument products, based on ascending or descending normal series. We give the relevant definitions.

([3]) This relation is well known in the general situation of universal algebras. See, for example, [25].

Let γ be an ordinal number and M_γ the set (segment) of all ordinal numbers α with $\alpha \leqslant \gamma$. We assume, further, that with every nonlimit $\alpha \in M_\gamma$ there is associated a certain class of groups $\mathfrak{X}_\alpha$. We define the *upper product* of the classes $\mathfrak{X}_\alpha$ as the class $\mathfrak{X}$ consisting of all groups G that have an ascending normal series $[A_\alpha]$, with repetitions allowed, of length γ, and such that for every $\alpha < \gamma$ we have $A_{\alpha+1}/A_\alpha \in \mathfrak{X}_{\alpha+1}$. Similarly, on the basis of descending series, we define the *lower product* of classes $\mathfrak{X}_\alpha$. We can equally well speak of *invariant* or *characteristic* upper or lower products, if we start out from invariant or characteristic series, respectively. In general, these products differ from each other.

The following theorem holds [5].

THEOREM 1. *The invariant upper product of any set of radical classes is itself a radical class, and this product automatically turns out to be characteristic. The lower normal product of coradical classes is also a coradical class, and the product is characteristic.*

The upper normal product of radical classes also turns out to be a radical class, provided that the number of factors is finite. In general, this is not the case, and the corresponding property holds only for special radical classes (see §5).

Of course, we can also speak of upper and lower powers of a single class.

§2. Group-theoretical functions

1. *Functorials.* Every group has a center, a derived group, and many other characteristic subgroups. It is natural to regard such subgroups as values of certain special functions $\mathfrak{F}$ defined on the class of all groups or on a sublcass of it and associating with every group G (or every group in a given class K) one of its subgroups $\mathfrak{F}(G)$. Here the following condition of abstractness must be satisfied: $\mathfrak{F}(G^\varphi) = \mathfrak{F}(G)^\varphi$ for each isomorphism φ between G and G^φ. We call such abstract group-theoretical functions *functorials*. From the definitions it follows at once that if $\mathfrak{F}$ is a functorial, then in each group G its value $\mathfrak{F}(G)$ is a characteristic subgroup. Apart from functorials it is sometimes useful to consider also *cofunctorials*, which associate with a group a factor group of it.

Of course, abstract group-theoretical functions in implicit form have been studied for a long time, but only in the recent paper by Baer [9] have they been named explicitly and regarded as an object in their own right. Group-theoretical functions and their applications in the theory of radicals are treated in the author's paper [5].

2. *Operations with functorials.* In the system of all functorials there are some natural relations and operations. First of all, the order relation: if $\mathfrak{F}_1$ and $\mathfrak{F}_2$ are two functorials, then $\mathfrak{F}_1 < \mathfrak{F}_2$ if $\mathfrak{F}_1(G) \subset \mathfrak{F}_2(G)$ for every group G. Next, let $\mathfrak{F}_\alpha$, $\alpha \in I$, be a collection of functorials, where I is not necessarily a set. The *compositum* and *intersection* $\cup_\alpha \mathfrak{F}_\alpha$ and $\cap_\alpha \mathfrak{F}_\alpha$ are defined by the following rules: for every group G the subgroup $(\cup_\alpha \mathfrak{F}_\alpha)(G)$ is the compositum of the subgroups $\mathfrak{F}_\alpha(G)$, and $(\cap_\alpha \mathfrak{F}_\alpha)(G)$ is the intersection of all the $\mathfrak{F}_\alpha(G)$.

There are two multiplications, the *upper* and the *lower*. Let $\mathfrak{F}_1$ and $\mathfrak{F}_2$ be two functorials. By $\mathfrak{F}_1 \circ \mathfrak{F}_2$ we denote the function that associates with a group G its subgroup $\mathfrak{F}_1 \circ \mathfrak{F}_2(G)$, which is the complete inverse image in G of the subgroup $\mathfrak{F}_2(G/\mathfrak{F}_1(G))$ of $G/\mathfrak{F}_1(G)$. This is the upper product. The lower product $\mathfrak{F}_1 \wr \mathfrak{F}_2$ is defined by the rule $\mathfrak{F}_1 \wr \mathfrak{F}_2(G) = \mathfrak{F}_2(\mathfrak{F}_1(G))$. Both these multiplications are associative operations.

We can also speak of infinite upper and lower products. Let M_γ be a segment of ordinal numbers less than a given γ, and suppose that with every nonlimit number $\alpha \in M_\gamma$ a functorial $\mathfrak{F}_\alpha$ is associated. We define the product $\Pi^\circ(\gamma)\mathfrak{F}_\alpha$ by induction, having defined $\Pi^\circ(\beta)\mathfrak{F}_\alpha$ for all $\beta \leqslant \gamma$. We set $\Pi^\circ(1)\mathfrak{F}_\alpha = \mathfrak{F}_1$. Suppose that for all $\eta < \beta$ the products $\Pi^\circ(\eta)\mathfrak{F}_\alpha$ are already defined. If β is not a limit number, then we define $\Pi^\circ(\beta)\mathfrak{F}_\alpha = (\Pi^\circ(\beta - 1)\mathfrak{F}_\alpha) \circ \mathfrak{F}_\beta$. But if β is a limit number, then $\Pi^\circ(\beta)\mathfrak{F}_\alpha$ is defined as the compositum of all $\Pi^\circ(\eta)\mathfrak{F}_\alpha$ with $\eta < \beta$. Similarly, the lower product $\Pi^\wr(\gamma)\mathfrak{F}_\alpha$ is defined by means of intersections.

For a single functorial $\mathfrak{F}$ we can consider its upper and lower powers. They are denoted, respectively, by $\mathfrak{F}^{\circ\gamma}$ and $\mathfrak{F}^{\wr\gamma}$, where γ is an ordinal number. We also set

$$\bar{\mathfrak{F}} = \bigcup_\gamma \mathfrak{F}^{\circ\gamma}; \qquad \check{\mathfrak{F}} = \bigcap_\gamma \mathfrak{F}^{\circ\gamma},$$

where γ ranges over all transfinite numbers. Here we have $\bar{\mathfrak{F}} \circ \mathfrak{F} = \bar{\mathfrak{F}}$ and $\bar{\mathfrak{F}} \circ \bar{\mathfrak{F}} = \bar{\mathfrak{F}}$, and, correspondingly, $\check{\mathfrak{F}} \wr \mathfrak{F} = \check{\mathfrak{F}}$ and $\mathfrak{F} \wr \check{\mathfrak{F}} = \check{\mathfrak{F}}$. A functorial $\mathfrak{F}$ for which $\mathfrak{F} \circ \mathfrak{F} = \mathfrak{F}$ is called *upper idempotent*, and if $\mathfrak{F} \wr \mathfrak{F} = \mathfrak{F}$, *lower idempotent*.

We can point out a number of other, more specific operations. There is, for example, *commutation*: if $\mathfrak{F}_1$ and $\mathfrak{F}_2$ are two functorials, their commutator $[\mathfrak{F}_1, \mathfrak{F}_2]$ acts according to the rule

$$[\mathfrak{F}_1, \mathfrak{F}_2](G) = [\mathfrak{F}_1(G), \mathfrak{F}_2(G)].$$

We can regard the system of all functorials as a certain algebraic system, although it is not a set. In particular, we can speak of the *upper* and *lower semigroups of functorials*, and in these semigroups we can single out various interesting subsemigroups. In [5] some of the simplest relations for the operations treated above are derived.

3. *Some restrictions imposed on functorials.* Various restrictions on functorials arise from an analysis of their behavior with respect to the operations defined above. Thus, we have already selected the upper and lower idempotents. We can also characterize functorials by their behavior in various group-theoretical constructions. For example, we can select functorials that commute with direct products. Below we name some restrictions to be applied in the theory of radicals and coradicals.

A functorial $\mathfrak{F}$ is called H-closed if $\mathfrak{F}(G)^\varphi = \mathfrak{F}(G^\varphi)$ for any epimorphism φ: $G \to G^\varphi$. If here only the inclusion $\mathfrak{F}(G)^\varphi \subset \mathfrak{F}(G^\varphi)$ holds, then $\mathfrak{F}$ is called *left H-closed*, and if $\mathfrak{F}(G)^\varphi \supset \mathfrak{F}(G^\varphi)$, then $\mathfrak{F}$ is called *right H-closed*. Right H-closure does not occur in our discussion; therefore, we call left H-closed functorials *strong*, for want of a better word. H-closed functorials are also called *verbal*, because they always select verbal subgroups with respect to some variety (see the next section).

Now let ρ be an abstract relation between a group and a subgroup of it. If A is a subgroup of G, then $A\rho G$ means that ρ holds for A and G. We say that a functorial $\mathfrak{F}$ is ρ-*compatible* if $A\rho G$ implies that $\mathfrak{F}(A) \subset \mathfrak{F}(G)$. We call $\mathfrak{F}$ ρ-*hereditary* if from $A\rho G$ it follows that $\mathfrak{F}(G) \cap A \subset \mathfrak{F}(A)$. If $\mathfrak{F}$ is simultaneously ρ-compatible and ρ-hereditary, then $A\rho G$ implies that $\mathfrak{F}(A) = \mathfrak{F}(G) \cap A$.

The relations of "being a normal subgroup", "being a subnormal subgroup" and "being an ascendant subgroup" are applied very frequently in the theory of radicals. It is also useful to consider the following type of relations. Let μ be an infinite cardinal. We define the relation $\rho = \rho(\mu)$ by the following rule: $A\rho G$ if A is a term of an ascending normal series of G of length less than the given μ. Of course, for every concrete group the relation of being ascendant coincides with some $\rho(\mu)$. If μ is a countable cardinal, then $\rho(\mu)$ is subnormality.

It is natural to single out functorials that associate with every group a fully invariant subgroup of it. In this context we note the following simple proposition.

If $\mathfrak{F}$ is a strong functorial and compatible with subgroups, then $\mathfrak{F}(G)$ is always a fully invariant subgroup of G.

For if φ is an endomorphism of G, then under the given conditions we have $\mathfrak{F}(G)^{\varphi} \subset \mathfrak{F}(G^{\varphi}) \subset \mathfrak{F}(G)$.

Every verbal satisfies this condition.

§3. Radicals and coradicals

1. *Two special relationships between functorials and classes of groups.* Every functorial $\mathfrak{F}$ singles out in a natural way two classes of groups: $\mathfrak{F}'$ and $\mathfrak{F}^*$. The class $\mathfrak{F}'$ consists of all groups G for which $\mathfrak{F}(G) = G$. We call such groups $\mathfrak{F}$-*complete*. $\mathfrak{F}^*$ is the class of all G for which $\mathfrak{F}(G) = E$. We call the groups of this class $\mathfrak{F}$-*simple*.

On the other hand, with every abstract class of groups $\mathfrak{X}$ two functorials are associated: $\mathfrak{X}'$ and $\mathfrak{X}^*$. The first of these selects in every group G the subgroup $\mathfrak{X}'(G)$ (also denoted by $\mathfrak{X}(G)$) that is generated by all normal $\mathfrak{X}$-subgroups of G. The second functorial singles out in G the subgroup $\mathfrak{X}^*(G)$ which is the intersection of all normal co-$\mathfrak{X}$-subgroups of G; that is, those A for which $G/A \in \mathfrak{X}$.

The meaning of the notation $\mathfrak{F}''$, $\mathfrak{F}^{**}$, $\mathfrak{X}''$ and $\mathfrak{X}^{**}$ is clear, and the equalities $\mathfrak{F} = \mathfrak{F}''$, $\mathfrak{X} = \mathfrak{X}''$, $\mathfrak{F} = \mathfrak{F}^{**}$ and $\mathfrak{X} = \mathfrak{X}^{**}$ express here that the corresponding mappings are bijections.

We say that a class of groups $\mathfrak{X}$ is a *preradical class* if in every group G its subgroup $\mathfrak{X}(G)$ is an $\mathfrak{X}$-subgroup. $\mathfrak{X}$ is called *copreradical* if always $\mathfrak{X}^*(G)$ is a co-$\mathfrak{X}$-subgroup. It is easy to see that a class $\mathfrak{X}$ is preradical if and only if it is closed under the operator R, and copreradical if it is CoR-closed. On the other hand, it is not hard to see that $X'' = \mathsf{R}\mathfrak{X}$ and $\mathfrak{X}^{**} = \mathsf{CoR}\,\mathfrak{X}$. Thus, the equality $\mathfrak{X} = \mathfrak{X}''$ is equivalent to $\mathfrak{X}$ being a preradical class, and $\mathfrak{X} = \mathfrak{X}^{**}$ means that $\mathfrak{X}$ is a copreradical class.

Now we make the parallel definition for functorials. A functorial $\mathfrak{F}$ is called *preradical* if 1) $\mathfrak{F}$ is lower idempotent and 2) $\mathfrak{F}$ is compatible with normal subgroups. $\mathfrak{F}$ is called *copreradical* if 1) $\mathfrak{F}$ is upper idempotent and 2) $\mathfrak{F}$ is a strong functorial. It is not hard to observe that both the first and second conditions are self-dual.

The following proposition is proved in [9] and [5].

PROPOSITION 2. *The equality $\mathfrak{F} = \mathfrak{F}''$ holds if and only if $\mathfrak{F}$ is preradical. The equality $\mathfrak{F} = \mathfrak{F}^{**}$ is equivalent to the fact that $\mathfrak{F}$ is copreradical. If $\mathfrak{F}$ is preradical, then $\mathfrak{F}'$ is a preradical class, and if $\mathfrak{X}$ is a preradical class, then $\mathfrak{X}'$ is preradical. If $\mathfrak{F}$ is copreradical, then $\mathfrak{F}^*$ is a copreradical class, and if $\mathfrak{X}$ is any class, then $\mathfrak{X}^*$ is copreradical.*

Thus, the symbol $'$ establishes a one-to-one correspondence between preradicals and preradical classes, and $*$ gives a similar relationship between copreradicals and copreradical classes.

Now we define $r\mathfrak{F} = \mathfrak{F}''$ and $\mathsf{Cor}\, \mathfrak{F} = \mathfrak{F}^{**}$, and we list some properties of these operators on functorials [5].

We always have $r\mathfrak{F} > \check{\mathfrak{F}}$ and $\mathsf{Cor}\, \mathfrak{F} < \bar{\mathfrak{F}}$. Consequently, if $\mathfrak{F}$ is lower idempotent, then $r\mathfrak{F} > \mathfrak{F}$, and if $\mathfrak{F}$ is upper idempotent, then $\mathsf{Cor}\, \mathfrak{F} < \mathfrak{F}$. If $\mathfrak{F}$ is a strong functorial, then $\mathsf{Cor}\, \mathfrak{F} = \bar{\mathfrak{F}}$, and if $\mathfrak{F}$ is compatible with normal subgroups, then $r\mathfrak{F} = \check{\mathfrak{F}}$. Furthermore, if $\mathfrak{F}$ is a strong functorial, so is $r\mathfrak{F}$, and if $\mathfrak{F}$ is compatible with normal subgroups, so is $\mathsf{Cor}\, \mathfrak{F}$.

The operator Cor coincides with its square. Therefore, for any $\mathfrak{F}$ the functorial $\mathsf{Cor}\, \mathfrak{F}$ is already a copreradical. However, this copreradical need not be covered by $\mathfrak{F}$. But if $\mathfrak{F}$ is upper idempotent, then $\mathsf{Cor}\, \mathfrak{F}$ is the maximal copreradical in $\mathfrak{F}$: it contains every copreradical less than $\mathfrak{F}$.

The operator r does not coincide with its square, and its powers can be defined in the obvious fashion, including infinite powers. These powers become stationary at the first limit ordinal. We write $\bar{r} = r^\omega$. The operator $\bar{r}$ assigns to every functorial $\mathfrak{F}$ the preradical $\bar{r}\mathfrak{F}$. If $\mathfrak{F}$ is lower idempotent, then $\bar{r}\mathfrak{F}$ is the minimal preradical covering $\mathfrak{F}$.

We note the following relations, which can be verified immediately:

$$(r\mathfrak{F})' = \mathsf{R}\mathfrak{F}'; \qquad (\mathsf{R}\mathfrak{X})' = r\mathfrak{X}';$$
$$(\mathsf{Cor}\, \mathfrak{F})^* = \mathsf{CoR}\, \mathfrak{F}^*; \qquad (\mathsf{CoR}\, \mathfrak{X})^* = \mathsf{Cor}\, \mathfrak{X}^*.$$

2. *Definition of radicals and coradicals. Simplest relations.* In accordance with previous remarks, a radical class is a preradical class with the additional condition that it is closed under homomorphisms. Similarly, a coradical class is a copreradical class that is hereditary with respect to normal subgroups. Going over to the language of functorials, we define the radical as a functorial $\mathfrak{F}$ satisfying the two conditions for a preradical and also the condition that $\mathfrak{F}$ is strong. Similarly, in the definition of a coradical $\mathfrak{F}$ we add to the two conditions for a copreradical the condition that $\mathfrak{F}$ is compatible with normal subgroups.

Now we observe that the third condition in the definition of a radical occurs as the second condition in the definition of a copreradical, and vice versa, the third condition in the definition of a coradical is also the second condition in the definition of a preradical. It is easy to see that here ′ is also a one-to-one correspondence between radical classes and radicals, and * between coradical classes and coradicals.

Next we recall that a radical $\mathfrak{F}$ is *hereditary* (with respect to normal subgroups) if from the fact that A is a normal subgroup of G it follows that $\mathfrak{F}(A) = \mathfrak{F}(G) \cap A$. This condition, as is easy to see, implies the first condition in the definition of a preradical. $\mathfrak{F}$ is hereditary if and only if the corresponding radical class is hereditary. The dual condition for coradical classes indicates that these classes are closed under homomorphisms, and this is equivalent to the corresponding coradical being H-closed. It turns out that H-closure of an arbitrary functorial already indicates that the functorial is coradical.

A functorial $\mathfrak{F}$ is called *verbal* if $\mathfrak{F} = \mathfrak{X}^*$, where $\mathfrak{X}$ is a variety. Clearly, every verbal is a coradical, is H-closed, and is compatible with subgroups. The following proposition holds (see, for example, [5]).

PROPOSITION 3. *A functorial $\mathfrak{F}$ is verbal if and only if it is H-closed. A class $\mathfrak{X}$ is a variety if and only if $\mathfrak{X} = \mathfrak{X}^{**}$ and $\mathfrak{X}^*$ is H-closed.*

In this way we also establish a duality between hereditary radicals and verbals. The duality between hereditary radical classes and varieties is a result of the definition of these classes in the operator language (about duality, see also §3.5).

Next, we remark that side by side with varieties and verbals it is sometimes useful to consider prevarieties and preverbals, which are defined as follows.

A class $\mathfrak{X}$ is called a *prevariety* if it is closed under the operator $\{\mathsf{S}, \mathsf{C}\}$ or, what is equivalent, under $\{\mathsf{S}, \mathsf{CoR}\}$. Thus, if $\mathfrak{X}$ is any class of groups, then $\mathfrak{X}_1 = \mathsf{CoR}\,\mathfrak{X}$ is a copreradical class, $\mathfrak{X}_2 = \mathsf{S}\mathfrak{X}_1$ is a prevariety (and at the same time a coradical class), and $\mathfrak{X}_3 = \mathsf{H}\mathfrak{X}_2$ is a variety. It is not hard to observe that every quasi-variety of groups [10] is always a prevariety, but not conversely. See also [32], where another name is used for prevarieties.

A functorial $\mathfrak{F}$ is called a *preverbal* if it is a copreradical and compatible with subgroups (but not only with normal subgroups, as in the definition of a coradical). $\mathfrak{F}$ is a preverbal if and only if the corresponding $\mathfrak{F}^*$ is a prevariety. From the preceding remarks it follows that in every group a preverbal singles out a fully invariant subgroup.

A radical $\mathfrak{F}$ is said to be *strict* if it is also upper idempotent. We observe here that the term "strict radical" is not very felicitous, because it is already used in another situation in [3]. A coradical $\mathfrak{F}$ is called *strict* if it is lower idempotent. The condition for a radical to be strict is equivalent to the corresponding class being closed under extensions; that is, to being a radical class in the axiomatics of Kuroš [3]. $\mathfrak{F}$ is a strict coradical if and only if $\mathfrak{F}^*$ is a coradical class that is closed under extensions, which is equivalent to semisimplicity of this class in the sense of

Kuroš. The following assertion follows immediately from the definitions: *if $\mathfrak{F}$ is a strict radical, then it is at the same time a strict coradical; if $\mathfrak{F}$ is a strict coradical, then it is also a strict radical.* $\mathfrak{F}'$ is always a strict radical class, and $\mathfrak{F}^*$, a semisimple class.

The following proposition is connected with the operators r and Cor introduced earlier [5].

PROPOSITION 4. *If $\mathfrak{F}$ is a strong functorial, then $\bar{r}\mathfrak{F}$ is radical. If $\mathfrak{F}$ is compatible with normal subgroups, then* Cor $\mathfrak{F}$ *is coradical.*

3. *Compatibility and heredity.* Compatibility and heredity carry a certain duality: if $\mathfrak{F}$ is a functorial and ρ a relation between a group and one of its subgroups, then ρ-heredity of $\mathfrak{F}$ implies heredity with respect to ρ-subgroups of the class $\mathfrak{F}'$, and ρ-compatibility of $\mathfrak{F}$ implies heredity with respect to ρ-subgroups of the class $\mathfrak{F}^*$. Now we give a criterion for ρ-compatibility of a radical.

We define an operator R_ρ on classes of groups: a group G belongs to the class $\mathsf{R}_\rho\mathfrak{X}$ if and only if it is generated by its ρ-subgroups in $\mathfrak{X}$. In parallel, we define an operator r_ρ on functorials: $\mathsf{r}_\rho\mathfrak{F}(G)$ is the subgroup of G generated by all $\mathfrak{F}$-complete ρ-subgroups in G. Obviously, $(\mathsf{r}_\rho\mathfrak{F})' = \mathsf{R}_\rho\mathfrak{F}'$.

Now we list a few restrictions that can be imposed on a relation ρ:

1) Every normal subgroup of a group is a ρ-subgroup of it.

2) If $A\rho G$ and B is a normal subgroup of G containing A, then $A\rho B$.

3) If $B\rho G$ and A is a characteristic subgroup of B, then $A\rho G$.

Every relation ρ satisfying these three conditions is called *normal*.

A relation ρ is called *strictly normal* if it also satisfies:

4) If $A\rho G$ and $\varphi\colon G \to G^\varphi$ is an epimorphism, then $A^\varphi\rho G^\varphi$.

Furthermore, it often becomes necessary to assume ρ to be transitive.

It is easy to note that if ρ satisfies 1), 2) and is transitive, then $(\mathsf{R}_\rho\mathfrak{X})'(G)$ is the subgroup of G generated by all its ρ-subgroups in $\mathfrak{X}$, and we have $(\mathsf{R}_\rho\mathfrak{X})' = \mathsf{r}_\rho\mathfrak{X}'$. For such a ρ it is also obvious that R_ρ is a closure operator and that $\mathsf{r}_\rho^2 = \mathsf{r}_\rho$.

PROPOSITION 5 [5]. *Let ρ be normal. A class of groups $\mathfrak{X}$ is preradical with a ρ-compatible $\mathfrak{X}'$ if and only $\mathsf{R}_\rho\mathfrak{X} = \mathfrak{X}$. A preradical $\mathfrak{F}$ is ρ-compatible if and only if $\mathsf{r}_\rho\mathfrak{F} = \mathfrak{F}$. If ρ is transitive, then for any $\mathfrak{F}$ the functorial $\mathsf{r}_\rho\mathfrak{F}$ is a ρ-compatible preradical. If ρ is also strictly normal and $\mathfrak{F}$ is a strong functorial, then $\mathsf{r}_\rho\mathfrak{F}$ is a ρ-compatible radical.*

For a strict normal ρ the following is also true: if $\mathfrak{F}$ is a ρ-compatible functorial, then Cor $\mathfrak{F}$ is likewise ρ-compatible, and (Cor $\mathfrak{F}$)* is a hereditary coradical class with respect to ρ-subgroups.

For a normal ρ it is not hard to verify that the intersection of any collection of ρ-hereditary radicals is itself a ρ-hereditary radical. Hence for every radical $\mathfrak{F}$ we can indicate the minimal ρ-hereditary radical containing $\mathfrak{F}$.

4. *Multiplication of radicals and coradicals.* Earlier on we defined the upper and lower products of functorials. We now note the following theorem [5], which is parallel to Theorem 1.

THEOREM 2. *The upper product of any set of radicals is itself a radical. The lower product of any set of coradicals is itself a coradical. If $\mathfrak{F}$ is radical, so is $\overline{\overline{\mathfrak{F}}}$; if $\mathfrak{F}$ is coradical, so is $\check{\mathfrak{F}}$.*

This theorem is proved by verifying the corresponding properties of functorials, and there is no need to go over to classes of groups. Furthermore, it can be verified that under minimal restrictions on ρ the upper product of ρ-compatible radicals is itself ρ-compatible, and if all the factors are ρ-hereditary, so is their product.

From the characterization of verbals contained in Proposition 3 there follows at once the known fact that the lower product of finitely many verbals is itself a verbal. Similarly, the commutator of verbals is a verbal, and it is easy to see that the commutator of two coradicals is a coradical.

Multiplications of classes and functorials are connected in a natural manner. We denote by $\Pi^{\sigma}(\gamma)\mathfrak{X}_{\alpha}$ the upper invariant product of radical classes, and by $\Pi^{\rho}(\gamma)\mathfrak{X}_{\alpha}$ the lower normal product of coradical classes. Then

$$(\Pi^{\sigma}(\gamma)\mathfrak{X}_{\alpha})' = \Pi^{\sigma}(\gamma)\mathfrak{X}'_{\alpha} \quad \text{and} \quad (\Pi^{\rho}(\gamma)\mathfrak{X}_{\alpha})^* = \Pi^{\rho}(\gamma)\mathfrak{X}^*_{\alpha}.$$

The first of these is also true for finite normal products, and, if all the factors are special radicals, for infinite normal products. Here a radical $\mathfrak{X}$ is called *special* if $\mathfrak{X}'$ is compatible with ascendant subgroups.

Taking into account that the upper and lower multiplications of functorials are always associative and applying the formulas just noted, we see that we can speak of the semigroups of radical and coradical classes. These semigroups turn out to be isomorphic, respectively, to the semigroup of radicals (under upper multiplication) and that of coradicals (under lower multiplication). The second of these semigroups contains the subsemigroup of verbals, which is isomorphic to the semigroup of varieties.

5. *Radicals and coradicals as functors.* So far, radicals and coradicals have stood out by two qualities: as certain classes of groups and as certain group-theoretical functions. Now we regard them as functors in suitable categories, and here the duality between them becomes particularly clear: it reduces to a reversal of arrows.

Let K be a category and Φ a one-argument covariant functor from this category into itself. We call Φ *radical* on K if the following conditions hold:

1. Together with Φ there is also defined a function f that associates with every object A a monomorphism $f(A)$: $\Phi(A) \to A$.

2. If we are given a morphism μ: $A \to B$, $\alpha = f(A)$ and $\beta = f(B)$, then the following diagram is commutative:

$$
\begin{array}{ccc}
\Phi(A) & \xrightarrow{\Phi(\mu)} & \Phi(B) \\
\alpha \downarrow & & \downarrow \beta \\
A & \xrightarrow{\mu} & B
\end{array}.
$$

3. $\Phi^2 = \Phi$.

The first two conditions indicate that Φ is a subfunctor of the identity functor. If they are replaced by duality, we arrive at the notion of a coradical. Thus, Φ is a coradical if it satisfies the following conditions:

1. Together with Φ there is also defined a function f that associates with every object A an epimorphism $f(A)$: $A \to \Phi(A)$.

2. If we are given a morphism μ: $A \to B$, $\alpha = f(A)$ and $\beta = f(B)$, then the following diagram is commutative:

$$
\begin{array}{ccc}
A & \xrightarrow{\;\mu\;} & B \\
\alpha \downarrow & & \downarrow \beta \\
\Phi(A) & \xrightarrow[\Phi(\mu)]{} & \Phi(B)
\end{array}
$$

3. $\Phi^2 = \Phi$.

Next, we call a homomorphism of groups μ: $A \to B$ *accessible* if its image is a subnormal subgroup of B. It is easy to see that the class of all groups, considered together with accessible homomorphisms, is a category, say, K.

Let Φ be a radical in K. For every group A we denote by $\mathfrak{F}(A)$ the image of the monomorphism $f(A)$. Then $\mathfrak{F}$ turns out to be a radical in the class of all groups. On the other hand, to every radical $\mathfrak{F}$ there corresponds, uniquely to within equivalence of functors, a radical Φ in K.

Now let Φ be a coradical in K. For every group A we denote by $\mathfrak{F}(A)$ the kernel of the epimorphism $f(A)$. Then $\mathfrak{F}$ is a coradical in the class of all groups, and here we also have a one-to-one correspondence.

Now let us assume that K is some category in which the concept of an exact sequence makes sense. Then we define a radical Φ on K to be *hereditary* if the functor Φ is left-exact. if Φ is a coradical and right-exact, we also call it *hereditary*.

Now let K be a category of groups, as above, with the natural concept of an exact sequence. Then the following proposition holds [5].

PROPOSITION 6. 1. *A radical Φ in K is hereditary if and only if the corresponding $\mathfrak{F}$ is a hereditary radical in the class of all groups.*

2. *A coradical Φ in K is hereditary if and only if the corresponding $\mathfrak{F}$ is verbal.*

From this it also follows that if Φ is a hereditary radical in K, then its images range over a hereditary radical class of groups, and if Φ is a hereditary coradical, then its images form a variety of groups.

6. *Additional remarks.* In all our discussions radicals and coradicals were defined on the class of all groups—they are *absolute* radicals and coradicals. However, it is clear that we could start out from a subclass K of the class of all groups that is closed, say, under homomorphisms and normal subgroups. The whole theory could be referred to this class K. Here, all the functorials must be defined on K, and the action of operators on classes must be restricted to K (on this, see [5]).

Now we make a few remarks about generalizations. Many of the above constructions connected with radicals can be generalized to Ω-groups, and for the theory of coradicals we can even go further and carry it over to universal algebras. Then almost everything connected with the correspondence * is preserved, provided that we have in mind that in a universal algebra a functorial fixes one of its congruences. What does not go over, at least not immediately, are the facts connected with the multiplication of coradicals. Even in Ω-groups they do not transfer at once, because here characteristic features of the ideals selected by functorials are altogether absent.

This is also the reason for difficulties in the theory of radicals in Ω-groups. We note that in the case of Ω-groups condition 2) in the definition of a radical $\mathfrak{F}$ must be replaced by the following condition: *if A is an $\mathfrak{F}$-perfect ideal of an Ω-group G, then $A \subset \mathfrak{F}(G)$.* With this modification of the definition, the simplest facts connected with the correspondence ' go over to Ω-groups, as is not hard to verify.

Various generalizations of the theory of radicals occur as a result of placing it within category theory. Although there are quite a number of papers on this topic, the necessary clarity for a category-theoretical treatment of the radical has not yet been achieved by any means.

§4. The radical closure of classes of groups

1. *The operator $\overline{\mathsf{R}}$ and related operators.* We recall that the operator R associates with a class $\mathfrak{X}$ of groups the class $\mathsf{R}\mathfrak{X}$ consisting of the groups that are generated by their normal $\mathfrak{X}$-subgroups. The following theorem characterizes the closure of this operator.

THEOREM 3. *R is a restriction operator. $\overline{\mathsf{R}} = \mathsf{R}^{\omega+1}$, and this is sharp. Furthermore, the closure $\overline{\mathsf{R}}$ can be defined as the operator that assigns to a class $\mathfrak{X}$ the class of all groups that are generated by their subnormal $\mathfrak{X}$-subgroups.*

Let us consider a more general situation. let ρ be a normal relation between a group and one of its subgroups, and $\bar{\rho}$ its closure with respect to transitivity. We claim that $\overline{\mathsf{R}}_{\rho} = \mathsf{R}_{\bar{\rho}} = \mathsf{R}_{\rho}^{\omega+1}$. To begin with, we verify that $\overline{\mathsf{R}}_{\rho} = \overline{\mathsf{R}}_{\bar{\rho}}$. Indeed, let $G \in \mathsf{R}_{\bar{\rho}}\mathfrak{X}$, $A\bar{\rho}G$ and $A \in \mathfrak{X}$. Of course, $\overline{\mathsf{R}}_{\bar{\rho}}\mathfrak{X}$ is a preradical class satisfying the condition of ρ-compatibility. Even $\bar{\rho}$-compatibility holds; therefore, $A = (\mathsf{R}_{\rho}\mathfrak{X})(A) \subset (\overline{\mathsf{R}}_{\rho}\mathfrak{X})(G)$. Since G is generated by these A, we see that $G = (\overline{\mathsf{R}}_{\rho}\mathfrak{X})(G)$, and this means that $G \in \overline{\mathsf{R}}_{\rho}(\mathfrak{X})$. Thus, $\mathsf{R}_{\bar{\rho}} \leqslant \overline{\mathsf{R}}_{\rho}$. On the other hand, it is evident that $\mathsf{R}_{\rho} \leqslant \mathsf{R}_{\bar{\rho}}$, and, since on the right there stands a closure operator, $\overline{\mathsf{R}}_{\rho} \leqslant \overline{\mathsf{R}}_{\bar{\rho}}$.

Next we show that $\mathsf{R}_{\bar{\rho}} = \mathsf{R}_{\rho}^{\omega+1}$. If $\mathsf{R}_{\rho} = \mathsf{R}$, we can utilize here standard arguments (see, for example, Ščukin [11]); in the general case we need a few auxiliary facts.

Let ρ be a relation and $\mathfrak{X}$ a class of groups. By $\mathfrak{X}^{\rho}$ we denote the functorial that singles out in every group G its subgroups generated by all the ρ-subgroups of G contained in $\mathfrak{X}$. When ρ is normal, it is obvious that $\mathfrak{X}^{\rho}(G) \in \mathsf{R}_{\rho}\mathfrak{X}$, $(\mathfrak{X}^{\rho})' = \mathsf{R}_{\rho}\mathfrak{X}$, and if $\mathsf{R}_\rho \mathfrak{X} = \mathfrak{X}$, then $\mathfrak{X}^\rho = \mathfrak{X}'$. We write $A\rho^n G$ if G has an ascending series

$$A\rho A_1 \rho \cdots \rho A_{n-1} \rho G. \tag{$*$}$$

Of course, if $A\rho G$, then $\mathfrak{X}^{\rho^{n-1}}(A) \subset \mathfrak{X}\rho^n(G)$.

Next we show that

$$\mathfrak{X}^{\rho^n}(G) = \left(\mathsf{R}_\rho^{n-1}\mathfrak{X}\right)^\rho(G).$$

This relation is evident for $n = 1$, and we assume that it is proved for all $m < n$. Let $A \in \mathfrak{X}$ and $A\rho^n G$. For this A we choose the series $(*)$ and let $B_{n-1} = \mathfrak{X}^{\rho^{n-1}}(A_{n-1})$. Then $A \subset B_{n-1}\rho G$. From the inductive hypothesis we have $B_{n-1} = (\mathsf{R}_\rho^{n-2}\mathfrak{X})^\rho(A_{n-1}) \in \mathsf{R}_\rho^{n-1}\mathfrak{X}$. Consequently, $B_{n-1} \subset (\mathsf{R}_\rho^{n-1}\mathfrak{X})^\rho(G)$. Since all these B_{n-1} generate $\mathfrak{X}^{\rho^n}(G)$, we have $\mathfrak{X}^{\rho^n}(G) \subset (\mathsf{R}_\rho^{n-1}\mathfrak{X})^\rho(G)$. Observing now that $\mathsf{R}_\rho^{n-1}\mathfrak{X} \subset \mathsf{R}_{\rho^{n-1}}\mathfrak{X}$,[4] we obtain the reverse inclusion $(\mathsf{R}_\rho^{n-1}\mathfrak{X})^\rho(G) \subset \mathfrak{X}^{\rho^n}(G)$.

Now let $G \in \mathsf{R}_{\bar\rho}\mathfrak{X}$. Then G has an ascending series of characteristic subgroups $\mathfrak{X}^{\rho^n}(G)$ whose union is the whole group. From the relation proved above it follows that $\mathfrak{X}^{\rho^n}(G) \in \mathsf{R}_\rho^n\mathfrak{X}$. Thus $G \in \mathsf{R}\cdot\mathsf{R}_\rho^\omega\mathfrak{X}$, and hence $\mathsf{R}_{\bar\rho} \leqslant \mathsf{R}\cdot\mathsf{R}_0^\omega$. Since $\mathsf{R}_{\bar\rho} = \bar{\mathsf{R}}_\rho$, we have $\mathsf{R}_\rho^{\omega+1} \leqslant \mathsf{R}_{\bar\rho}$. So we have proved that $\mathsf{R}_{\bar\rho} = \mathsf{R}\cdot\mathsf{R}_\rho^\omega = \mathsf{R}_\rho^{\omega+1}$.

The fact that the inequalities in the theorem are sharp has recently been proved by Rips [12]. He has shown that the upper bound is attained in the class $\mathfrak{N}$ of nilpotent groups. There we have the strictly ascending sequence of classes

$$\mathfrak{N} \subset \mathsf{R}\mathfrak{N} \subset \mathsf{R}^2\mathfrak{N} \subset \cdots \subset \mathsf{R}^n\mathfrak{N} \subset \cdots,$$

and the required property follows from this.

Observe that the groups of the class $\mathsf{R}\mathfrak{N}$ are the Fitting groups, and that $\bar{\mathsf{R}}\mathfrak{N}$ is the class of all *Baer groups*. The fact that $\mathsf{R}\mathfrak{N} \neq \bar{\mathsf{R}}\mathfrak{N}$ was established somewhat earlier by Dark [13]. In fact, Dark solved an old problem by proving that there exist Baer groups without Abelian normal subgroups.

Now let $\mathfrak{X}$ be a class of groups and $\mathsf{H}\mathfrak{X} = \mathfrak{X}$. Then $\bar{\mathsf{R}}\mathfrak{X} = \bar{\mathsf{R}}^{\omega+1}\mathfrak{X}$ for such an $\mathfrak{X}$ is a radical class. It may happen, however, that $\bar{\mathsf{R}}\mathfrak{X} = \mathsf{R}^\alpha\mathfrak{X}$ for some $\alpha < \omega + 1$. Let $\tau = \tau(\mathfrak{X})$ be the first among the numbers α for which $\mathsf{R}^\alpha\mathfrak{X}$ is a radical class. This τ, $0 \leqslant \tau \leqslant \omega + 1$, cannot be equal to ω, as is easy to see. If $\tau = 0$, it means that $\mathfrak{X}$ already is a radical class, and here we have a certain characteristic measure for the deviation from being a radical class. We have seen that if $\mathfrak{X}$ is the class of all nilpotent groups, then τ takes the values $\omega + 1$. It is also known (see, for example, [14]) that if $\mathfrak{X}$ is a class of simple noncommutative groups, then always $\tau(\mathfrak{X}) = 1$. In various concrete cases the computation of τ can turn out to be an interesting problem. It is natural to raise the following question.

PROBLEM 1. *It is true that for every natural number n there exists an* H-*closed class of groups $\mathfrak{X}$ for which $\tau(\mathfrak{X}) = n$?*

We mention that we do not know a single case when τ is different from 0, 1 and $\omega + 1$.

---

[4] Actually, the equality $\mathsf{R}_\rho^n = \mathsf{R}_{\rho^n}$ holds, as follows from the formula to be proved.

A similar problem for another operator connected with strict radical closures was mentioned earlier by Kuroš (see our Problem 8).

Now we consider some operators closely related to R and $\overline{R}$. By $K\mathfrak{X}$ we denote the class of groups that are covered by their normal $\mathfrak{X}$-subgroups. Clearly $K < R$, and therefore $\overline{K} < \overline{R}$. The operator $\overline{K}$ was studied by Kontorovič in his important paper [15]. We call it the *Kontorovič operator*. In the case when $\mathfrak{X} = \mathfrak{A}$ is the class of all Abelian groups, the corresponding $\overline{K}\mathfrak{A}$ is the class of groups with a category. It is easy to see that $\overline{R}\mathfrak{A}$ is the class of all Baer groups, and since $\overline{K} < \overline{R}$, we see that every group with a category is a Baer group. In fact, the following theorem holds (Rips [31]).

THEOREM 4. *The class of groups with a category coincides with the class of Baer groups.*

Thus, Baer groups, which are important for the theory of radicals, could just as well be called Baer-Kontorovič groups. We have seen that the operators $\overline{K}$ and $\overline{R}$ act identically on the class of all Abelian groups. In [31] Rips has pointed out some other classes of groups on which $\overline{K}$ and $\overline{R}$ agree. This is the case for all finite groups, for all Noetherian groups, etc. In the same paper he considers the operator F, which is defined as follows: $F\mathfrak{X}$ is the class of groups in which every finite set of elements belongs to a normal $\mathfrak{X}$-subgroup. Of course, $F < K$. He discusses some classes of groups on which the action of $\overline{F}$ and $\overline{R}$ is identical, and he gives some bounds connected with F. For example, he shows that on the class of all Noetherian groups this operator is bounded, but its bound is substantially larger than the bound of R.

We also note that the action of the operators $\overline{K}$ and $\overline{F}$ on the class of finite groups are considered in the paper by Kalužnin [16], which is concerned with generalized locally normal groups.

We now define the following operator V: $G \in V\mathfrak{X}$ if every finite set of elements of G is contained in a subnormal $\mathfrak{X}$-subgroup. Of course, $F < V < \overline{R}$; and, since V is evidently a closure operator, $\overline{F} < V < \overline{R}$. Apparently the operators $\overline{F}$ and V are distinct; however, we have no proof for this. In the recent paper [19] by Roseblade and Stonehewer the following very useful theorem is proved.

THEOREM 5. *If $\mathfrak{X}$ is a hereditary class with respect to normal subgroups and if in any group the product of two normal $\mathfrak{X}$-subgroups is again an $\mathfrak{X}$-subgroup, then $\overline{R}\mathfrak{X} = V\mathfrak{X}$.*

This theorem was known previously [20] under the additional assumption that the groups in $\mathfrak{X}$ are Noetherian.

We denote by R_0 the operator that associates with a class $\mathfrak{X}$ the class of groups that can be generated by two normal $\mathfrak{X}$-subgroups. Of course, $\overline{R}_0 = R_0^\omega$. Next, let $R_1\mathfrak{X}$ be the class of groups that are unions of an ascending sequence of normal $\mathfrak{X}$-subgroups. It is easy to see that $\overline{R} = \{R_0, R_1\}$. Furthermore, the relation $\overline{R} = R_1 R^\omega = FR^\omega$ is evident.

2. *The radical and hereditary closure of classes.* As already mentioned, the operator Rad that associates with a class $\mathfrak{X}$ the minimal radical class containing $\mathfrak{X}$ has the following simple structure: $\mathsf{Rad} = \overline{\mathsf{R}}\mathsf{H}$. It is also easy to see that if $\mathfrak{X}$ is a preradical class, then the subclass $\mathfrak{X}_0$ of $\mathfrak{X}$ consisting of all groups in $\mathfrak{X}$ whose homomorphic images also lie in $\mathfrak{X}$ is a radical class. The transition from $\mathfrak{X}$ to $\mathfrak{X}_0$ is brought about by a class restriction operator.

The minimal hereditary radical class containing a given $\mathfrak{X}$ is $\{\mathsf{R}, \mathsf{H}, \mathsf{I}\}\mathfrak{X}$. We do not know whether $\{\mathsf{R}, \mathsf{H}, \mathsf{I}\}$ can be reduced to some finite product of $\overline{\mathsf{R}}$, H, and $\overline{\mathsf{I}}$. Apparently this is not the case, because the operators R and I interact badly. We only note the following simplification, which is often helpful.

PROPOSITION 7. $\{\mathsf{R}, \mathsf{H}, \mathsf{I}\} = \overline{\mathsf{R}}\{\mathsf{R}_0, \mathsf{H}, \mathsf{I}\}$.

To verify this formula it is enough to check that for any $\mathfrak{X}$ the class $\mathsf{R}\{\mathsf{R}_0, \mathsf{H}, \mathsf{I}\}\mathfrak{X}$ is closed under R, H, and I. That it is closed under R and H is obvious, and for I we obtain this by means of Theorem 5. We write $\mathfrak{X}_1 = \{\mathsf{R}_0, \mathsf{H}, \mathsf{I}\}\mathfrak{X}$. By Theorem 5, $\overline{\mathsf{R}}\mathfrak{X}_1 = \mathsf{V}\mathfrak{X}_1$, and the operator V interacts well with I: $\mathsf{IV} \leqslant \mathsf{VI}$. Therefore, the class $\mathsf{V}\mathfrak{X}_1$ is closed under I.

Next, let ρ be a relation between a group and one of its subgroups. We denote by S_ρ the operator that associates with a class $\mathfrak{X}$ the class of all ρ-subgroups of groups in $\mathfrak{X}$. We can also speak of radical classes that are hereditary with respect to ρ-subgroups; therefore, it is desirable also to know the structure of the operator $\{\mathsf{R}, \mathsf{H}, \mathsf{S}_\rho\}$. It is easy to see that if ρ is such that $A\rho G$ implies $(A \cap B)\rho B$ for any subnormal subgroup B of G, then for such a ρ we have $\mathsf{S}_\rho\mathsf{V} < \mathsf{VS}_\rho$. If every normal subgroup is a ρ-subgroup, then, as above, we obtain the relation $\{\mathsf{R}, \mathsf{H}, \mathsf{S}_\rho\} = \overline{\mathsf{R}}\{\mathsf{R}_0, \mathsf{H}, \mathsf{S}_\rho\}$.

Let us consider applications to specific classes $\mathfrak{X}$. Let $\mathfrak{S}$ be the class of all soluble groups. Clearly, $\{\mathsf{R}_0, \mathsf{H}, \mathsf{S}\}\mathfrak{S} = \mathfrak{S}$. Therefore, the hereditary radical class relative to subgroups generated by $\mathfrak{S}$ is $\overline{\mathsf{R}}\mathfrak{S} = \mathsf{V}\mathfrak{S}$. All its groups are locally soluble. Since $\mathfrak{S}$ is the union of the classes $\mathfrak{A}^n$ for various n (where $\mathfrak{A}$ is the class of all Abelian groups), then by applying the relevant formula from §6.1 we obtain the inclusion $\overline{\mathsf{R}}\mathfrak{S} \subset (\overline{\mathsf{R}}\mathfrak{A})^\omega$. Here $\overline{\mathsf{R}}\mathfrak{A}$ is the class of Baer groups, and the power of a class is understood to be the upper invariant power.

Now let $\mathfrak{N}_\mu$ be the class of all groups whose cardinal is less than a given infinite cardinal μ. Of course, $\{\mathsf{R}_0, \mathsf{H}, \mathsf{S}\}\mathfrak{N}_\mu = \mathfrak{N}_\mu$, and hence $\{\mathsf{R}, \mathsf{H}, \mathsf{S}\}\mathfrak{N}_\mu = \overline{\mathsf{R}}\mathfrak{N}_\mu = \mathsf{V}\mathfrak{N}_\mu$. It is also clear that if $\mathfrak{M}$ is the class of Noetherian groups, then the hereditary radical class relative to subgroups generated by this class coincides with $\overline{\mathsf{R}}\mathfrak{M}$. Furthermore, according to [20], every group in $\overline{\mathsf{R}}\mathfrak{M}$ has a local system of subnormal Noetherian subgroups. Here we mention that from results in [19] it follows that if $\mathfrak{F}$ is the class of all finitely generated groups, then every group in the radical class $\overline{\mathsf{R}}\mathfrak{F}$ has a local system of finitely generated subnormal subgroups.

Now let $\mathfrak{X}$ be the class of all Artinian groups; that is, groups with minimal condition. Clearly, $\{\mathsf{R}_0, \mathsf{H}, \mathsf{S}\}\mathfrak{X} = \mathfrak{X}$, and therefore the corresponding hereditary radical is $\overline{\mathsf{R}}\mathfrak{X} = \mathsf{V}\mathfrak{X}$.

We fix the following notation: if A is a group, then $[A]$ denotes the class of all groups isomorphic to A.

PROPOSITION 8. *Let A be a Baer group, and let $\mathfrak{X} = \{R, H, I\}[A]$. If A is not a periodic group, then $\mathfrak{X}$ is the class of all Baer groups. If A is a periodic group and π the set of all prime divisors of the orders of its elements, then $\mathfrak{X}$ is the class of Baer π-groups. In both cases $\mathfrak{X}$ also turns out to be hereditary with respect to subgroups.*

PROOF. To begin with, suppose that A is not periodic. Then A has an infinite cyclic subgroup, say A_0. Since A is a Baer group, A_0 is subnormal in A; and since the groups in $\mathfrak{X}$ are hereditary with respect to normal subgroups, $A_0 \in \mathfrak{X}$. Now $\mathfrak{X}$ contains all homomorphic images of A_0; therefore, the class $\mathfrak{X}$ contains all cyclic groups. By applying to this class the operator $\overline{R}$ we obtain the class of all Baer groups. Hence, $\mathfrak{X}$ is the class of all Baer groups.

Now let A be periodic and π the relevant set of prime numbers. As above, we show that $\mathfrak{X}$ contains cyclic groups of prime order p_i for all $p_i \in \pi$. Let A_0 be one of these groups, of order p. We write $A_0^{(2)} = A_0 \wr A_0,\dots,A_0^{(n)} = A_0^{(n-1)} \wr A_0,\dots$. Then $A_0^{(n)}$ is nilpotent and is generated by its (subnormal) subgroups of order p. Consequently $A_0^{(n)} \in \mathfrak{X}$. Now $A_0^{(n)}$ contains as a subnormal subgroup a cyclic group of order p^n. We have shown that $\mathfrak{X}$ contains cyclic groups of order p^n for all n, with p ranging over the whole set π. Since every Baer π-group is generated by its subnormal cyclic p-subgroups with $p \in \pi$, it is now clear that $\mathfrak{X}$ is the class of all Baer π-groups.

Let us call two groups A and B *equivalent* if $\{R, H, I\}[A] = \{R, H, I\}[B]$. We have just seen that if A and B are nonperiodic Baer groups, then they are equivalent. If A and B are periodic Baer groups, they are equivalent if and only if they have the same set π. The investigation of criteria for radical equivalence of groups leads to various interesting problems. In particular, V. G. Viljacer and A. S. Hahutaišvili (in a paper being prepared for publication)* consider some criteria for radical equivalence of wreath products.

Many interesting problems are connected with the study of conditions similar to radical equivalence of classes of groups.

3. *The operator of special radical closure.* In applications to individual groups, especially in questions connected with infinite ascending normal series, an important role is played by special radicals; that is, radicals compatible with ascendant subgroups. If ρ is the property of a subgroup of being ascendant, then we denote the corresponding R_ρ by $\tilde{R}$. If $\mathfrak{X}$ is any class of groups, its special radical closure is the class $\tilde{R}H\mathfrak{X}$.

We denote by $\tilde{R}_1\mathfrak{X}$ the class of groups that are unions of an ascending normal sequence of $\mathfrak{X}$-subgroups.

THEOREM 6 [21]. $\tilde{R} = \{\tilde{R}_1, R_0\}$. *In particular, a radical class $\mathfrak{X}$ is special if and only if it is $\tilde{R}_1$-closed.*

---

* *Editor's note.* This paper appears never to have been published.

This theorem also contains the following assertion. *If $\mathsf{L}\mathfrak{X}$ is the class of all groups having a local system of $\mathfrak{X}$-subgroups, then every L-closed radical class is special.* For these strong special radical classes $\mathfrak{X}$ (under the additional condition of heredity) the following assertion holds [4]: *If A is a locally ascendant subgroup of G, then $\mathfrak{X}(A) \subset \mathfrak{X}(G)$.* Here, A is locally ascendant in G if G has a local system of subgroups containing A such that A is ascendant in each of them.

In §3.2 we defined the relation $\rho(\mu)$ for infinite cardinals μ. We denote the corresponding operator $\mathsf{R}_{\rho(\mu)}$ by R_μ. Of course, for every $\mathfrak{X}$ the class $\tilde{\mathsf{R}}\mathfrak{X}$ is the union of the $\mathsf{R}_\mu\mathfrak{X}$ for all μ, and if μ_0 is the first infinite cardinal, then $\mathsf{R}_{\mu_0}\mathfrak{X} = \bar{\mathsf{R}}\mathfrak{X}$.

It is also easy to see that always

$$\{\mathsf{R}_\mu, \mathsf{H}\} = \mathsf{R}_\mu\mathsf{H}.$$

The following theorem concerns the structure of the operators $\tilde{\mathsf{R}}$ and R_μ as applied to some special classes $\mathfrak{X}$.

THEOREM 7. *Suppose that all groups of the class $\mathfrak{X}$ are Noetherian, and that $\mathsf{R}_0\mathfrak{X} = \mathfrak{X}$ and $\mathsf{I}\mathfrak{X} = \mathfrak{X}$. Then $\tilde{\mathsf{R}}\mathfrak{X}$ is the class of groups having a local system of ascendant $\mathfrak{X}$-subgroups, and $\mathsf{R}_\mu\mathfrak{X}$ is the class of groups having a local system of $\mathfrak{X}$-subgroups that are subnormal in fewer than μ steps.*

This theorem immediately follows from the main result of [20]. From the same result it is also easy to deduce that in any group the compositum of two ascendant Noetherian subgroups is itself an ascendant Noetherian subgroup.

Like Theorem 5, Theorem 7 can be applied to construct hereditary special radical closures of classes of groups. However, the working of this theorem is restricted by the additional condition of being Noetherian. It would be interesting to investigate to what extent this condition can be weakened. On this topic we note a few problems.

PROBLEM 2. *Let $\mathfrak{X}$ be the class of groups that are extensions of Artinian groups by Noetherian groups. Is it true that the special radical class $\tilde{\mathsf{R}}\mathfrak{X}$ is hereditary with respect to subgroups?*

We mention here that it is known from [22] that the compositum of two ascendant Artinian subgroups in any group is itself an ascendant Artinian subgroup.

PROBLEM 3. *Let $\mathfrak{S}$ be the class of all soluble groups and $G \in \tilde{\mathsf{R}}\mathfrak{S}$. Is it true that every finite set of elements in G is contained in an ascendant soluble subgroup? Is it true perhaps that G is locally soluble?*

It is easy to see that $\tilde{\mathsf{R}}S \subset (\tilde{\mathsf{R}}A)^\omega$.

PROBLEM 4. *Let μ be an infinite cardinal and $\mathfrak{K}_\mu$ the class of all groups whose cardinal is less than μ. Is the radical class $\tilde{\mathsf{R}}\mathsf{K}_\mu$ hereditary with respect to subgroups?*

The following problems refer to the operators R_μ.

PROBLEM 5. *Do there exist classes of groups $\mathfrak{X}$ for which all the $\mathsf{R}_\mu\mathfrak{X}$ are distinct?*

More is required in the following problem.

PROBLEM 6. *Does there exist a class of groups $\mathfrak{X}$ such that for any two distinct cardinals μ_α and μ_β, $\mu_\alpha < \mu_\beta$, there is a group G such that $G \in \mathsf{R}_{\mu_\beta}\mathfrak{X}$ and $\mathsf{R}_{\mu_\alpha}\mathfrak{X}(G) = E$?*

It is possible that the last two problems already have an affirmative solution in the class of Abelian groups or that of finite groups. If $\mathfrak{A}$ is the class of all Abelian groups, then $\tilde{\mathsf{R}}\mathfrak{A}$ is the class of all *hypernilpotent* groups; that is, groups in which every cyclic subgroup is ascendant. This is a known fact, and also follows easily from Theorem 7. From the same theorem one deduces almost immediately that $\mathsf{R}_\mu\mathfrak{A}$ is the class of groups in which subgroups are subnormal in fewer than μ steps: μ-subnormal. We say that the groups of the class $\mathsf{R}_\mu\mathfrak{A}$ are *μ-hypernilpotent*. If G is any group, then $\tilde{\mathsf{R}}\mathfrak{A}(G)$ is the set of all ascendant elements, and $\mathsf{R}_\mu\mathfrak{A}(G)$ the set of all μ-subnormal elements of G. Here an element is μ-subnormal (ascendant) if the cyclic subgroup generated by it is.

Problem 6, which was mentioned above in its application to the class of Abelian groups, can now be formulated as follows.

PROBLEM 7. *Let μ_α and μ_β be two distinct infinite cardinals with $\mu_\alpha < \mu_\beta$. Does there exist a group G such that all its elements are μ_β-subnormal and G has no μ_α-subnormal elements?*

4. Strict radicals. As already mentioned, these are radicals that are idempotent in the semigroup of all radicals or, what is the same, radical classes that are closed under extensions. Let $\mathsf{E}\mathfrak{X}$ ($\mathsf{Ext}\,\mathfrak{X}$) be the class $\mathfrak{X}\mathfrak{X}$ of all groups that are extensions of $\mathfrak{X}$-groups by $\mathfrak{X}$-groups. The closure $\bar{\mathsf{E}}$ of this operator E associates with a class $\mathfrak{X}$ the class of all groups having finite normal series with $\mathfrak{X}$-factors. Of course, $\bar{\mathsf{E}} = \mathsf{E}^\omega$. It is also easy to see that $\mathsf{H}\bar{\mathsf{E}} < \bar{\mathsf{E}}\mathsf{H}$ and $\mathsf{S}\bar{\mathsf{E}} < \bar{\mathsf{E}}\mathsf{S}$, and therefore, $\{\mathsf{H}, \mathsf{E}\} = \bar{\mathsf{E}}\mathsf{H}$ and $\{\mathsf{S}, \mathsf{E}\} = \bar{\mathsf{E}}\mathsf{S}$.

Strict radical closure is brought about by the operator $\{\mathsf{R}, \mathsf{H}, \mathsf{E}\}$. To represent this and some similar operators in explicit form, we introduce some further operators.

J: $\mathsf{J}\mathfrak{X}$ is the class of groups having an ascending invariant series whose factors all belong to $\mathfrak{X}$.

N: $\mathsf{N}\mathfrak{X}$ is the class of groups having an ascending normal series with $\mathfrak{X}$-factors.

N_0: $\mathsf{N}_0\mathfrak{X}$ is the class of groups having an ascending normal series whose terms are all subnormal in the group with respect to this series and whose factors all belong to $\mathfrak{X}$.

N_μ: the class $\mathsf{N}_\mu\mathfrak{X}$ is defined just as $\mathsf{N}_0\mathfrak{X}$, but the condition of being subnormal is replaced by being μ-subnormal.

Evidently, N, N_0 and all the N_μ are closure operators, but J is not.

By means of these operators some standard theorems in the theory of radicals can be written down in the following formulas.

THEOREM 8.

$$1.\quad \{\mathsf{R}, \mathsf{H}, \mathsf{E}\} = \mathsf{J}^{\omega+1}\mathsf{H} = \mathsf{N}_0\mathsf{H} = \mathsf{J}\bar{\mathsf{R}}\mathsf{H}.$$

$$2.\quad \{\mathsf{R}, \mathsf{H}, \mathsf{E}\} = \mathsf{N}\mathsf{H} = \mathsf{J}\mathsf{R}\bar{\mathsf{H}}.$$

$$3.\quad \{\mathsf{R}_\mu, \mathsf{H}, \mathsf{E}\} = \mathsf{N}_\mu\mathsf{H} = \mathsf{J}\mathsf{R}_\mu\mathsf{H}.$$

Proofs of 1 and 2 can be found in Kuroš [3], Ščukin [11] and Plotkin [18]. 3 can be proved by the same methods. Of course, 2 describes a special strict radical closure, and 3 plays the analogous role for $\mathbf{R}_\mu$. The formulas referring to special radicals imply, in particular, the following important remark: *if $\mathfrak{X}$ is a special radical class, then* $\mathbf{N}\mathfrak{X} = \mathbf{J}\mathfrak{X}$.

Observe that the $\omega + 1$ in 1 is sharp. This follows from a result of Kovács and Neumann [23], who have shown that if $\mathfrak{A}$ is the class of all Abelian groups, then the sequence of classes $\mathfrak{A} \subset \mathbf{J}\mathfrak{A} \subset \mathbf{J}^2\mathfrak{A} \subset \cdots$ is strictly ascending. Our assertion about the exponent being sharp follows at once from this. On the other hand, the following problem due to Kuroš seems to be unsolved.

PROBLEM 8. *Does there exist for every n a class of groups $\mathfrak{X}$ such that $\mathbf{J}^n\mathbf{H}\mathfrak{X}$ is a strict radical class, but $\mathbf{J}^{n-1}\mathbf{H}\mathfrak{X}$ is not?*

In contrast to **Rad**, the operator of strict radical closure interacts well with operators connected with conditions of heredity. As a rule, if the original class $\mathfrak{X}$ is hereditary, so is its closure. We cannot dwell on this in detail, and only mention the following proposition, which is also connected with heredity.

Suppose that ρ is a relation between a group and one of its subgroups, that it is transitive, preserved under epimorphisms and under taking complete inverse images of epimorphisms, and suppose also that from $A\rho G$ it follows that $(A \cap B)\rho B$ for every subgroup B of G. Then for each such ρ we have

PROPOSITION 9. *Let $\mathfrak{X}$ be a strict radical class. Denote by $\mathfrak{X}_\rho$ the class of all groups that lie in $\mathfrak{X}$ together with all their ρ-subgroups. Then $\mathfrak{X}_\rho$ is a ρ-hereditary strict radical class. If $\mathfrak{X}$ is special, so is $\mathfrak{X}_\rho$.*

The proof reduces to a simple verification of the conditions, and we omit it.

§5. Coradical closure of classes

1. *Some further operators.* We lay down some notation. If $\mathfrak{X}$ is a class of groups we denote by $\daleth\, \mathfrak{X}$ the class of all groups in which there are no nontrivial subnormal $\mathfrak{X}$-subgroups. By $\llcorner\, \mathfrak{X}$ we denote the class of all groups for which no nontrivial homomorphic image is an $\mathfrak{X}$-group. In other words, $G \in \llcorner\, \mathfrak{X}$ if $\mathfrak{X}^*(G) = G$. Here we are also dealing with operators on classes of groups; however, these operators are not monotone.

According to the definition, a group G belongs to the class $\llcorner\, \daleth\, \mathfrak{X}$ if and only if every nontrivial homomorphic image of it has a nontrivial subnormal $\mathfrak{X}$-subgroup. If the class $\mathfrak{X}$ is closed under homomorphisms, we arrive at the equality $\llcorner\, \daleth\, \mathfrak{X} = \mathbf{N}_0\mathfrak{X}$. Thus, we observe that $\llcorner\, \daleth\, \mathfrak{X} = \mathfrak{X}$ together with $\mathbf{H}\mathfrak{X} = \mathfrak{X}$ indicates that $\mathfrak{X}$ is a strict radical class.

Let us analyze the class $\daleth\, \llcorner\, \mathfrak{X}$ in a similar way. The inclusion $G \in \daleth\, \llcorner\, \mathfrak{X}$ means that every nontrivial subnormal subgroup of $\mathfrak{X}$ has a nontrivial homomorphic image belonging to $\mathfrak{X}$. It is easy to see that under this condition G has a descending invariant series whose factors are all residually $\mathfrak{X}$-groups.

Now we introduce another two operators.

Co J: **Co J**$\mathfrak{X}$ is the class of groups having a descending invariant series with $\mathfrak{X}$-factors.

Co N: Co N$\mathfrak{X}$ is the class of groups having a descending normal series with $\mathfrak{X}$-factors.

It is easy to verify the following relation.

PROPOSITION 10. *If $\mathfrak{X}$ is a hereditary class with respect to normal subgroups* ($I\mathfrak{X} = \mathfrak{X}$), *then*

$$\daleth \sqcup \mathfrak{X} = \mathsf{Co\,J Co\,R}\mathfrak{X}.$$

A class $\mathfrak{X}$ satisfying the conditions $I\mathfrak{X} = \mathfrak{X}$ and $\mathsf{Co\,J Co\,R}\mathfrak{X} = \mathfrak{X}$ is semisimple and strictly coradical. Thus, we see that semisimple classes can also be described as those for which $I\mathfrak{X} = \mathfrak{X}$ and $\daleth \sqcup \mathfrak{X} = \mathfrak{X}$.

2. *The coradical and strict coradical closure of classes.* Coradical closure of classes is effected by the operator $\{\mathsf{Co\,R}, \mathsf{I}\}$. We have already seen that $\{\mathsf{Co\,R}, \mathsf{I}\} = \mathsf{Co\,R\bar{I}}$. This operator is dual to that of radical closure. The operator dual to that of hereditary radical closure is $\{\mathsf{Co\,R}, \mathsf{I}, \mathsf{H}\}$. This last operator, as already mentioned, is the same as $\mathsf{HCo\,R}$, which associates with every class of groups the variety generated by this class.

We now go over to strict coradical closure. The following theorem can be deduced from results of Kuroš [3] and Ch'ang Wang-hao [24].

THEOREM 9. $\{\mathsf{Co\,R}, \mathsf{I}, \mathsf{E}\} = \mathsf{Co\,N\bar{I}} = \mathsf{Co\,J Co\,R\bar{I}} = (\mathsf{Co\,J})^2\bar{\mathsf{I}}$.

From this theorem it follows, in particular, that if $\mathfrak{X}$ is a coradical class, then $\mathsf{Co\,N}\mathfrak{X} = \mathsf{Co\,J}\mathfrak{X}$.

THEOREM 10 (see Kuroš [3]). *If $H\mathfrak{X} = \mathfrak{X}$, then $\daleth\,\mathfrak{X}$ is a strict coradical class. If $I\mathfrak{X} = \mathfrak{X}$, then $\mathfrak{X}$ is a strict radical class.*

Furthermore, we have already seen that if $\mathfrak{X}$ is a strict radical class, then $\sqcup \daleth\, \mathfrak{X} = \mathfrak{X}$, and if $\mathfrak{X}$ is a strict coradical class, then $\daleth \sqcup \mathfrak{X} = \mathfrak{X}$.

We supplement this theorem with the following easily verifiable proposition.

PROPOSITION 11. *If $\mathfrak{X}$ is a special radical class, then $\daleth\,\mathfrak{X}$ is hereditary with respect to ascendant subgroups. If $\mathfrak{X}$ is hereditary with respect to ascendant subgroups, then $\sqcup\,\mathfrak{X}$ is a special strict radical class.*

In this proposition we can also start out from a general relation ρ between a group and one of its subgroups and connect in a similar fashion ρ-compatibility of a radical class with ρ-heredity of a semisimple class.

Theorem 10 establishes a one-to-one correspondence between semisimple and strict radical classes, and special strict radical classes are in one-to-one correspondence with semisimple classes that are hereditary with respect to subnormal subgroups.

3. *Simultaneously radical and coradical classes.* We prove the following proposition.

PROPOSITION 12. *If $\mathfrak{X}$ is at the same time a radical and coradical class, then $\mathfrak{X}$ is the class of all groups.*

PROOF. From the conditions it follows that $\mathfrak{X}$ is a variety (because $\mathfrak{X}$ is closed under the operators H and Co R), and that $\mathfrak{X}$ contains a cyclic group of prime order p. As in Proposition 8, we can establish by means of wreath products that $\mathfrak{X}$ contains all cyclic p-groups. The infinite cyclic group is residually a cyclic p-group; therefore, $\mathfrak{X}$ contains an infinite cyclic group. But then $\mathfrak{X}$ also contains the whole class of Baer groups, and in particular, all nilpotent groups. By means of the operator Co R we can include in $\mathfrak{X}$ all free groups. Applying the H-closure, we obtain the required property.

We note that this proposition ceases to be true if in all arguments the original class is taken to be not the class of all groups, but a subclass of it. For example, it is false in the class of all Abelian groups.

Closely related to the preceding analysis is the interesting problem of specifying radical classes by formulas of logical languages. From Proposition 12 it follows that if a radical class is at the same time a variety, then it is the class of all groups. The situation is preserved if the varieties are replaced here by quasi-varieties. More generally, we can claim that *if a radical class of groups is determined by universal formulas of LPC, then it is the class of all groups.* To prove the last assertion we have to recall the following facts from the theory of models [25]. A class of groups determined by universal formulas of LPC is one that can be determined by certain formulas of LPC and is hereditary with respect to subgroups. It is also known that if a class of groups can be axiomatized by formulas of LPC, then it is closed under Cartesian products whenever it is closed under direct products. It now remains to apply Proposition 12.

It is natural to raise the following question.

PROBLEM 9. *Do there exist nontrivial radical classes of groups that are determined by formulas of LPC?*

Coradical classes with this property, of course, exist.

§6. Operations with radical and coradical classes

1. *The semigroup of radical classes.* We have already mentioned that the system of all radical classes is an associative system closed under multiplication. We call it a semigroup, although it is not a set. The elements of this semigroup are also called radicals, where we have in mind the presence of an isomorphism. In passing, we deduce three formulas which have a bearing on the semigroup property of radicals:

$$H[\mathfrak{X}_1\mathfrak{X}_2] \subset (H\mathfrak{X}_1)(H\mathfrak{X}_2); \qquad I[\mathfrak{X}_1\mathfrak{X}_2] \subset (I\mathfrak{X}_1)(I\mathfrak{X}_2),$$

and, if $H\mathfrak{X}_1 = \mathfrak{X}_1$ and $H\mathfrak{X}_2 = \mathfrak{X}_2$, then

$$\overline{R}[\mathfrak{X}_1\mathfrak{X}_2] \subset (\overline{R}\mathfrak{X}_1)(\overline{R}\mathfrak{X}_2).$$

These formulas are evident, and it is not hard to see that all inclusions are strict.

One of the possible ways of classifying radicals is to treat them from the point of view of their behavior in the semigroup of all radicals. But this semigroup itself seems to be an object worthy of attention. In contrast to the semigroup of varieties, it is not free. It has idempotents, namely strict radicals; not every radical can be represented as a product of indecomposable ones; and a number of relations arise because of the presence of infinite products. All these deviations from freeness hold equally well in the semigroup of all hereditary radicals. The semigroup of hereditary radicals is discussed in [6], in which some free systems of radicals are singled out, and some criteria for indecomposability of radicals are discussed. For example, *every hereditary radical generated by an arbitrary class of soluble groups is an indecomposable element in the semigroup of all hereditary radicals*. A similar property is possessed by the *small radicals*; that is, radicals generated by a set of groups. Vovsi [26] has shown that *every hereditary radical that is not idempotent is free and even absolutely free*: all its powers, including the infinite ones, are distinct. In other words, if $\mathfrak{X}$ is a nonstrict hereditary radical, then for no ordinary number α is $\mathfrak{X}^{\alpha}$ strict. The dual proposition holds for prevarieties. This was proved by L. M. Martynov, independently of Vovsi. Vovsi has also noted that a strict hereditary radical cannot be embedded nontrivially in a product of hereditary radicals. Consequently, if from the semigroup of all hereditary radicals the idempotent ones are removed, then the remaining part is also a semigroup, and all its elements are free.

In this semigroup of free radicals there are relationships, and we do not know whether the classification of all these relationships is a genuine problem. On the other hand, the following is.

PROBLEM 10. *Are there nontrivial relationships in the semigroup generated by indecomposable hereditary radicals?*

The problem of classifying all indecomposable radicals appears altogether non-genuine. However, one can raise this problem in certain special cases. We mention the following problem.

PROBLEM 11. *To investigate the indecomposable hereditary radicals in the class of locally finite p-groups.*

Here, too, a complete classification is apparently non-genuine. But one can hope to select interesting series.

As already mentioned, the dual object to a hereditary radical is a variety, so that to a certain extent the semigroup of hereditary radicals is dual to that of varieties. The semigroup of varieties, in its turn, is contained in the semigroup of all coradicals. Nothing interesting is known so far about this latter semigroup.

2. *Unions and intersections of radicals.* Let $[\mathfrak{X}_{\alpha}]$ be a collection of classes of groups. It is easy to verify the following formula:

$$\overline{\mathsf{R}}\left[\bigcup_{\alpha} \mathfrak{X}_{\alpha}\right] = \mathsf{R}\left[\bigcup_{\alpha} \overline{\mathsf{R}}\mathfrak{X}_{\alpha}\right].$$

If here all the $\mathfrak{X}_{\alpha}$ are preradical classes, then

$$\bigcup_{\alpha} \mathfrak{X}_{\alpha}' = \left(\mathsf{R}\left[\bigcup_{\alpha} \mathfrak{X}_{\alpha}\right]\right)' = \left(\bigcup_{\alpha} \mathfrak{X}_{\alpha}\right)'.$$

It is also easy to see that if all the $\mathfrak{X}_{\alpha}$ are radical classes, then their compositum $\mathsf{R}[\bigcup_{\alpha} \mathfrak{X}_{\alpha}]$ also is a radical class. The intersection of radical classes is iself a radical class, and the intersection of hereditary radical classes a hereditary radical class. If all the $\mathfrak{X}_\alpha$ are hereditary radicals, then $(\cap_\alpha \mathfrak{X}_\alpha)' = \cap_\alpha \mathfrak{X}'_\alpha$.

Unfortunately, heredity is lost under taking composita of radical classes.

We now list a few easily verifiable facts on the connection between composita and intersection on the one hand, and multiplication on the other. There is left distributivity:

$$\mathfrak{X}\left(\mathsf{R}\left[\bigcup_\alpha \mathfrak{X}_\alpha\right]\right) = \mathsf{R}\left[\bigcup_\alpha \mathfrak{X}\mathfrak{X}_\alpha\right].$$

The corresponding right distributivity does not hold, but only the inclusion

$$\mathsf{R}\left[\bigcup_\alpha \mathfrak{X}_\alpha\mathfrak{X}\right] \subset \mathsf{R}\left[\bigcup_\alpha \mathfrak{X}_\alpha\right]\mathfrak{X}.$$

For intersections we have $\mathfrak{X}(\cap_\alpha \mathfrak{X}_\alpha) = \cap_\alpha \mathfrak{X}\mathfrak{X}_\alpha$. Right distributivity does not hold here. However, if α ranges over a finite set and all the relevant radicals are hereditary with respect to subgroups, then right distributivity holds.

More on coradicals. If all the $\mathfrak{X}_\alpha$ are coradical, so is $\cap_\alpha \mathfrak{X}_\alpha$. The union of coradical classes need not be coradical: the operator $\mathsf{Co\,R}$ has yet to be applied to this union. We note the following formula:

$$\left(\mathsf{Co\,R}\left[\bigcup_\alpha \mathfrak{X}_\alpha\right]\right)^* = \bigcap_\alpha \mathfrak{X}_\alpha^* = \left(\bigcup_\alpha \mathfrak{X}_\alpha\right)^*.$$

Having in mind the connection with multiplication, we note that right distributivity holds:

$$\mathsf{Co\,R}\left[\bigcup_\alpha \mathfrak{X}_\alpha\right]\mathfrak{X} = \mathsf{Co\,R}\left[\bigcup_\alpha \mathfrak{X}_\alpha\mathfrak{X}\right].$$

For radical classes in a similar situation we had left, but not right distributivity. Here (for coradicals) there is no left distributivity. For intersections we have $(\cap_\alpha \mathfrak{X}_\alpha)\mathfrak{X} = \cap_\alpha(\mathfrak{X}_\alpha\mathfrak{X})$.

In conclusion we mention that many results on operations with radicals are proved by means of wreath products. In this context, some rules for computation of radicals in wreath products are derived in [6] and [26].

§7. Radicals in individual groups

1. *Computation of the radical in a group.* In a number of cases a radical in a specific group can be characterized as the set of group elements with a definite property. For example, the Baer radical, the hypernilpotent radical, and the locally nilpotent radical have such characterizations. A similar characterization can be given for all the μ-hypernilpotent radicals.

In [20] there is a general device for obtaining characterizations of radicals in terms of elements. Observe also that some standard facts on certain sets of elements of a group with certain given properties being subgroups can be obtained via radicals. There are several theorems on an elementwise characterization of radicals in groups with definite properties. For example, if a group has an ascending normal series with locally Noetherian factors, then the locally nilpotent coincides with the set of all left Engel elements.

For every concrete group we can raise the problem of describing all its radicals, the values of all radical functorials in it. In connection with this we mention that it is not hard to derive necessary and sufficient conditions under which a subgroup turns out to be radical with respect to some radical class. Similar conditions can be derived also for coradicals.

2. *Radicals and structural schemes.* Various structural schemes are known in group theory that help in the investigation of specific groups. Radicals and coradicals often are part of such schemes. As a rule, radicals work in those cases when we are concerned with ascending normal series, and coradicals are applied in problems connected with descending normal series.

Many applications are connected with a study of finiteness conditions. One often uses not only absolute radicals, but also radicals of given classes: in the presence of finiteness conditions some classes turn out to be radical that are not, in general, radical.

A major role is played in applications by the fact that a radical often bounds a group, i.e. contains its centralizer. There are several theorems on conditions when this takes place. The list of references contains a number of papers on the relevant problems.

In addition to the problems already mentioned, we list some others.

PROBLEM 12. *Does there exist a radical class of groups, wider than the class of locally nilpotent groups, and consisting of groups having a central system?*

PROBLEM 13. *Is it true that for every radical $\mathfrak{X}$ the following property holds: if G is a locally nilpotent torsion-free group, then $\mathfrak{X}(G)$ is an isolated subgroup of G?*

PROBLEM 14. *Is the locally nilpotent radical generated by its torsion-free groups?*

PROBLEM 15. *Can a group have more than one, but finitely many, maximal locally soluble normal subgroups?*

Infinitely many such normal subgroups can exist [28].

The following problems are concerned with radicals in some specific classes of groups. It is not hard to describe all hereditary radicals among the nilpotent groups of class n (for any n) and in the class of all nilpotent groups [29]. A similar problem can be raised for the case when nilpotency is replaced by solubility. Here we confine ourselves to the following problems.

PROBLEM 16. *To describe all hereditary radicals in the class of metabelian groups. The same in the class of polycyclic groups.*

In [3] Kuroš has shown how to find all hereditary strict radicals in the class of finite groups. We mention the following problem.

PROBLEM 17. *To find all hereditary radicals in the class of finite groups.*

PROBLEM 18. *To study the semigroup of hereditary radicals in the class of finite groups.*

Let Σ denote the semigroup just mentioned. It is not hard to see that Σ has the cardinal of the continuum. It is natural to raise the problem of describing the indecomposable elements of Σ. Here we mention a few special problems in this direction. Let $\mathfrak{S}_n$ be the element of Σ that is generated by the class of all n-step soluble (finite) groups, and $\mathfrak{N}$ the class of all finite nilpotent groups. Since $\mathfrak{N}$ is a radical class (the Fitting radical), we see that $\mathfrak{S}_n \subset \mathfrak{N}^n$. Apparently, the inclusion is strict. If this is so, then the question arises whether all the $\mathfrak{S}_n$ are indecomposable. This question is linked with an affirmative answer to the similar question for radicals in the class of all groups. We call a radical in Σ *small* if it is generated by a single group. Small radicals are, in particular, those consisting of all finite p-groups for every fixed p. Every such radical is idempotent. On the other hand, it appears plausible that if a small radical is not one of those mentioned, then it is indecomposable. We call a radical *simple* if it is generated by a collection of non-Abelian simple groups. A simple nil-radical is one generated by a collection of simple Abelian groups. It can be shown (A. S. Hahutaišvili) that a product of simple radicals cannot be small. However, it is not known whether the similar assertion for products of simple nil-radicals that are not idempotent is true or not. We mention the following problem: *Is it true that a subradical of a small radical is also small*?

There is hardly any research on radicals in finite groups, and it seems to us that an interesting circle of problems could arise here. Before formulating the following problem, which is due to P. Hall, we make a few remarks on radicals that arise in classes of simple groups.

Let $\mathfrak{P}$ be an abstract class of simple groups, and $\mathfrak{X}$ the class of all groups not in $\mathfrak{P}$. Then $\mathfrak{X}$ is a preradical, but not a radical class. It is not hard to show that $\mathfrak{X}$ is a special preradical class if and only if all the groups in $\mathfrak{P}$ are strictly simple; that is, do not contain proper ascendant subgroups. Next, let $\mathfrak{X}_1$ be the class of all groups that lie in $\mathfrak{X}$ together with all their homomorphic images. Clearly, $\mathfrak{X}_1 = \lrcorner \mathfrak{P}$, and this is a strict radical class. If all the groups in $\mathfrak{P}$ are simple, $\mathfrak{X}_1$ is a special radical.

We also introduce the class $\mathfrak{X}_2$ consisting of all groups that lie in $\mathfrak{X}_1$ together with all their subgroups. This is a strict radical class and hereditary with respect to subgroups.

Now we assume that the original class $\mathfrak{P}$ consists of all absolutely simple groups (non-Abelian groups without nontrivial normal systems). The corresponding $\mathfrak{X}_2$ is then denoted by $\mathfrak{H}$. It is easy to see that the class $\mathfrak{H}$ consists of all groups G such that if $B \subset G$ and $A \lhd B$, then the group B/A has a soluble normal system (is an SN-group). $\mathfrak{H}$ contains all simple non-Abelian groups. To exclude them we take for the original $\mathfrak{P}$ the class of all non-Abelian simple groups and denote by $\mathfrak{H}_0$ the corresponding $\mathfrak{X}_2$. Clearly $\mathfrak{H}_0 < \mathfrak{H}$. The class $\mathfrak{H}_0$ can also be defined in the following way: we take the class of all SN-groups and then choose in this class all groups for which no homomorphic image is a simple non-Abelian group, and finally we choose in the latter class all the groups lying in it together with all their subgroups. Then we come to the class $\mathfrak{H}_0$.

It is easy to see that $\mathfrak{H}$ is closed under the operator L, and therefore is a special radical class. Now we state the following problem from [7].

PROBLEM 19. *Is $\mathfrak{H}_0$ a special radical class*?

P. Hall's radical classes $\mathfrak{H}$ and $\mathfrak{H}_0$ are wider than radicals of soluble type. For example, $\mathfrak{H}$ contains all groups having an ascending normal series with locally radical factors. However, it is not known whether all such groups lie in $\mathfrak{H}_0$. In this connection we mention that, in answer to a question of P. Hall, Hartley and Stonehewer have recently proved in [30] that all locally radical groups belong to $\mathfrak{H}_0$.

Here, as in [21], radical groups are those having an ascending normal series with locally nilpotent factors. These groups form a special radical class, which is, however, not closed under the operator L.

<h2 style="text-align:center">BIBLIOGRAPHY</h2>

1. A. G. Kuroš, *Radicals of rings and algebras*, Mat. Sb. **33** (**75**) (1953), 13–26. (Russian)

2. S. A. Amitsur, *A general theory of radicals*. I, II, III, Amer. J. Math. **74** (1952), 774–786; **76** (1954), 100–125, 126–136.

3. A. G. Kuroš, *Radicals in group theory*, Sibirsk. Mat. Ž. **3** (1962), 912–931. (Russian)

4. B. I. Plotkin, *Groups of automorphisms of algebraic systems*, "Nauka", Moscow, 1966; English transl., Wolters-Noordhoff, Groningen, 1972.

5. ______, *The functorials, radicals and coradicals in groups*, Ural. Gos. Univ. Mat. Zap. **7** (1969/70), tetrad' 3, 150–182. (Russian)

6. ______, *The semigroup of radical classes of groups*, Sibirsk. Mat. Ž. **10** (1969), 1091–1108; English transl. in Siberian Math. J. **10** (1969).

7. Philip Hall, *On non-strictly simple groups*, Proc. Cambridge Philos. Soc. **59** (1963), 531–553.

8. S. R. Kogalovskiĭ, *On Birkhoff's theorem*, Uspehi Mat. Nauk **20** (1965), no. 5 (125), 206–207. (Russian)

9. Reinhold Baer, *Group theoretical properties and functions*, Colloq. Math. **14** (1966), 285–328.

10. A. I. Mal'cev, *Subdirect products of models*, Dokl. Akad. Nauk SSSR **109** (1956), 264–266; English tansl. in A. I. Mal'cev, *The metamathematics of algebraic systems. Collected papers*: 1936–1967, North-Holland, Amsterdam, 1971.

11. K. K. Ščukin, *On the theory of radicals in groups*, Sibirsk. Mat. Ž. **3** (1962), 932–942. (Russian)

12. I. A. Rips, *Two propositions on Baer groups*, Dokl. Akad. Nauk SSSR **186** (1969), 264–267; English transl. in Soviet Math. Dokl. **10** (1969).

13. R. S. Dark, *A prime Baer group*, Math. Z. **105** (1968), 294–298.

14. B. I. Plotkin, *Radical and semisimple groups*, Trudy Moskov. Mat. Obšč. **6** (1957), 299–336. (Russian)

15. P. G. Kontorovič, *Invariantly covered groups*. II, Mat. Sb. **28** (**70**) (1951), 79–88. (Russian)

16. L. A. Kalužnin,* *Locally normal groups of higher categories*, Algebra and Math. Logic: Studies in Algebra, Izdat. Kiev. Univ., Kiev, 1966, pp. 62–71. (Russian)

17. A. G. Kuroš, *Theory of groups*, 3rd ed., "Nauka", Moscow, 1967; English transl. of 2nd ed., vols. I, II, Chelsea, New York, 1955, 1956, 1960.

18. B. I. Plotkin, *Generalized solvable and generalized nilpotent groups*, Uspehi Mat. Nauk **13** (1958), no. 4 (82), 89–172; English transl. in Amer.. Math. Soc. Transl. (2) **17** (1961).

19. J. E. Roseblade and S. E. Stonehewer, *Subjunctive and locally coalescent classes of groups*, J. Algebra **8** (1968), 423–435.

20. B. I. Plotkin and Š. S. Kemhadze, *A scheme for constructing radicals in groups*, Sibirsk. Mat. Ž. **6** (1965), 1197–1201. (Russian)

21. B. I. Plotkin, *Radical groups*, Mat. Sb. **37** (**79**) (1955), 507–526; English transl. in Amer. Math. Soc. Transl. (2) **17** (1961).

22. Derek J. S. Robinson, *On the theory of subnormal subgroups*, Math. Z. **89** (1965), 30–51.

---

* *Editor's note*. While living in France (1938–1951) this author published a number of papers in French under the transliterated form "Kaloujnine" of his name.

23. L. G. Kovács and B. H. Neumann, *On the existence of Baur-soluble groups of arbitrary height*, Acta Sci. Math. (Szeged) **26** (1965), 143–144.

24. Čan Van-hao [Ch'ang Wang-hao], *On semisimple classes of groups*, Sibirsk. Mat. Ž. **3** (1962), 943–949. (Russian)

25. P. M. Cohn, *Universal algebra*, Harper & Row, New York, 1965.

26. S. V. Vorsi, *Absolute freedom of hereditary radicals*, Mat. Zametki (to appear).

27. B. I. Plotkin, *On semisimple groups*, Sibirsk. Mat. Ž. **9** (1968), 623–631; English transl. in Siberian Math. J. **9** (1968).

28. Gilbert Baumslag, L. G. Kovács and B. H. Neumann, *On products of normal subgroups*, Acta Sci. Math. (Szeged) **26** (1965), 145–147.

29. A. S. Hahutaišvili, *Hereditary radicals in the class of nilpotent groups*, Latvian Math. Yearbook, no. 7, "Zinatne", Riga, 1970, pp. 277–282. (Russian)

30. B. Hartley and S. E. Stonehewer, *On some questions of P. Hall concerning simple, generalized soluble groups*, J. London Math. Soc. **43** (1968), 739–744.

31. I. A. Rips, *On certain classes of groups of the type of groups with a category*, Ural. Gos. Univ. Mat. Zap. **7** (1969/70), tetrad' 3, 183–194. (Russian)

32. A. I. Mal'cev, *Algebraic systems*, "Nauka", Moscow, 1970; English transl., Springer-Verlag, Berlin and New York, 1973.

Translated by K. A. HIRSCH

Amer. Math. Soc. Transl.
(2) Vol. **119**, 1983

On Certain Nonassociative Nil Rings
and Algebraic Algebras*

A. I. ŠIRŠOV

§1. Introduction

In the works of Levitzki [7] and Jacobson [6], which are devoted to solving the Kuroš problem [1], it is shown that each associative algebra which is algebraic of bounded degree is locally finite, and each associative ring which is nil of bounded index is locally nilpotent. The Kuroš problem may be posed for any class of power-associative algebras [4]; however, already Lie algebras provide an example that this problem cannot have an affirmative solution for all power-associative algebras.

In this article we give a positive solution (again in the bounded case) of the analogous problems for special Jordan algebras and alternative algebras([1]) (Theorems 4 and 8). For nil rings we also get results generalizing the theorem of Levitzki for the associative case (Theorems 2 and 7).

§2. Preliminary results

We consider associative words generated by elements of some finite ordered set of symbols $R = \{a_1,\ldots,a_k\}$, where $a_i > a_j$ if $i > j$.

DEFINITION 1. We will call a word α a_k-*indecomposable* if it has the form $\alpha = a_k\ldots a_k a_{i_1} a_{i_2}\ldots a_{i_s}$ where $i_t \neq k$ $(t = 1,\ldots,s)$ and $s \geq 1$.

DEFINITION 2. A representation (if one exists) of the word β in the form of a product of a certain number of a_k-indecomposable words will be called an a_k-*decomposition* of the word β.

1980 *Mathematics Subject Classification.* Primary 17C50, 17D05, 16A68; Secondary 16A22.

([1]) *Translator's note.* In the original paper these results were obtained only for rings without 2-torsion and algebras over fields of characteristic $\neq 2$; by using quadratic instead of linear Jordan algebras, the proofs carry over almost verbatim to arbitrary rings of scalars. These changes have been made in the translation.

* Translation of Mat. Sb. **41 (83)** (1957), 381–394; MR **19**, 727. The translator has made changes, with the author's consent, that closely resemble the treatment to be found in the book *Rings that are nearly associative* ("Nauka" Moscow, 1978). Similar changes have been made in the following paper.

It is easy to see that a word β has an a_k-decomposition (moreover, a unique one) if and only if β begins with the symbol a_k and ends with a symbol different from a_k.

Below is given an example of an a_3-decomposition of a word on 3 symbols:

$$(a_3a_3a_2a_1a_1a_2a_1)(a_3a_1)(a_3a_3a_3a_1a_1a_2)(a_3a_1a_2).$$

In the set of all associative words on the elements of the set R we introduce a partial ordering: for words α and β, having the same length, we write $\alpha > \beta$ if this relation is satisfied in the lexicographic sense. We order the set T of all a_k-indecomposable words lexicographically; moreover, in the case where the word α is an initial word of β (i.e. $\beta = \alpha a_{i_1}i_{i_2} \cdots a_{i_m}$ with $i_t \neq k$, $t = 1,\ldots,m$) we write $\alpha > \beta$.

DEFINITION 3. We call an associative word γ *n-split* if it may be represented in the form of a product of n subwords such that under any nontrivial permutation of these subwords a new associative word is obtained which is strictly smaller than the word $\gamma(^2)$.

For example, the word $a_3a_1a_2a_2a_1a_1a_2a_1a_1a_1$ is 3-split and permits several 3-splittings:

$$(a_3a_1)(a_2a_2a_1a_1)(a_2a_1a_1a_1), \; (a_3a_1a_2)(a_2a_1a_1)(a_2a_1a_1a_1),$$
$$(a_3) \, (a_1a_2a_2a_1a_1a_2)(a_1a_1a_1), \text{ etc.}$$

A word which permits an a_k-decomposition may be considered as a word generated by elements of the set T, and in this case it makes sense to consider n-split words. In what follows, where this may give rise to misunderstanding we will speak of n_R-split and n_T-split words, stressing which set of symbols is being considered as the set of generators.

When we consider a word generated by elements of the set T, we will call it a *T-word* (in distinction to an *R-word*); in an analogous manner we will use the terms *R-length* and *T-length*.

LEMMA 1. *Let α be an associative T-word. Then, if it is n_T-split, it is also n_R-split.*

PROOF. Let $\alpha = \alpha_1 \cdots \alpha_n$ be an n_T-splitting of the word α; then α and $\alpha_1,\ldots,\alpha_n$ are words which permit a_k-decompositions. From Definition 3 it follows that $\alpha > \alpha_{i_1} \cdots \alpha_{i_n}$ in the sense of the set T whenever $(i_1,\ldots,i_n)$ is a nontrivial permutation of $(1,\ldots,n)$. It is easy to see that this inequality will also be valid in the sense of the set R. Therefore a given n_T-splitting will also be an n_R-splitting. Q.E.D.

LEMMA 2. *If a word α is $(n-1)_T$-split, then the word αa_k is n_R-split.*

$(^2)$ *Translator's note.* In general, γ is n-split if $\gamma = \gamma_1 \cdots \gamma_n$, where, for all permutations $\pi \neq 1$, $\gamma_1 \cdots \gamma_n > \gamma_{\pi(1)} \cdots \gamma_{\pi(n)}$. The importance of such words lies in lexicographic induction; if a polynomial identity on a ring guarantees $x_1 \cdots x_n = \sum_{\pi \neq 1} \lambda_\pi x_{\pi(1)} \cdots x_{\pi(n)}$ for all elements x_i, then an n-split word $\gamma_1 \cdots \gamma_n$ may be replaced by a linear combination of lexicographically lower words $\gamma_{\pi(1)} \cdots \gamma_{\pi(n)}$.

PROOF. From Lemma 1 follows the existence of the following $(n-1)_R$-splitting of the word α:

$$\alpha = \left(a_k a_{i_1} \cdots a'_{i_1}\right)\left(a_k a_{i_2} \cdots a'_{i_2}\right) \cdots \left(a_k a_{i_{n-1}} \cdots a'_{i_{n-1}}\right)$$

where $a, a' \in R$ and $a'_{i_t} \neq a_k$ for $t = 1, 2, \ldots, n-1$.

We will show that the word αa_k has the following n_R-splitting:

$$\alpha a_k = \left(a_k\right)\left(a_{i_1} \cdots a'_{i_1} a_k\right)\left(a_{i_2} \cdots a'_{i_2} a_k\right) \cdots \left(a_{i_{n-1}} \cdots a'_{i_{n-1}} a_k\right).$$

Indeed, any permutation of the factors of the word αa_k fixing the first factor a_k transforms the word αa_k into a word $\alpha' a_k$, where the word α' is obtained from α by means of some permutation of the factors of the given $(n-1)_T$-splitting of the word α. Therefore $\alpha > \alpha'$ and $\alpha a_k > \alpha' a_k$. If we consider a permutation which moves the symbol a_k from the first place, it is obvious that from its application to the word αa_k we obtain a word beginning with a strictly smaller number of symbols a_k compared to the word αa_k.   Q.E.D.

LEMMA 3. *For any three natural numbers k, s, n there exists a natural number $N(k, s, n)$ such that in any associative word of length $N(k, s, n)$ on k ordered symbols there will appear either s consecutive equal subwords, or else an n-split subword.*

PROOF. It is easily seen that natural numbers $N(k, s, 1) = 1$ and $N(1, s, n) = s$ satisfying the requirements of the lemma exist for all k, s and n. Let natural numbers k and n be given. We make an inductive assumption of the existence, for all natural numbers K and s, of natural numbers $N(k-1, s, n)$ and $N(K, s, n-1)$ satisfying the requirements of the lemma.

Consider an arbitrary associative word α of length

$$\left[s + N(k-1, s, n)\right]\left[N(k^{N(k-1,s,n)+s}, s, n-1) + 1\right]$$

on the elements of our set R. If at the beginning of the word α there stands some number of symbols a_i different from the symbol a_k, and their number is no smaller than the number $N(k-1, s, n)$, then the induction assumption is fulfilled in relation to the subword α' standing at the beginning of the word α and depending only on $k-1$ generators. Therefore we may assume that the length of such a word α', if it exists, is less than $N(k-1, s, n)$. At the end of the word α there may appear a subword $\alpha'' = a_k a_k \cdots a_k$. We assume that if such a word α'' exists, its length is less than s, since otherwise the conclusion of the lemma would be satisfied. Deleting the words α' and α'', if they exist, we obtain a subword α_1 beginning with a_k and ending with an a_i for $i \neq k$ whose length is larger than the number $[s + N(k-1, s, n)]N(k^{N(k-1,s,n)+s}, s, n-1)$.

Performing an a_k-decomposition of the word α_1, we assume in addition that the length of each a_k-indecomposable word appearing in that a_k-decomposition is smaller than the number $s + N(k-1, s, n)$, since otherwise either s consecutive symbols a_k would appear in such a word or a subword of length $N(k-1, s, n)$ not containing the symbol a_k. It is easy to see that there exist no more than

$K^{N(k-1,s,n)+s}$ different a_k-indecomposable words under the indicated restriction on their length. The word α_1 may be considered as a T-word. Since its T-length is strictly larger than $N(k^{N(k-1,s,n)+s}, s, n-1)$, either s consecutive equal subwords will appear in the word α_1 or else an $(n-1)_T$-split subword β.

If the second supposition is fulfilled, then, in view of the strict inequality for the length of the word α_1, we have the right to assume that the subword β is followed by the symbol a_k. By Lemma 2 the subword βa_k is n_R-split. In this case, and clearly in case the first supposition is fulfilled, the conclusion of the lemma is satisfied. Therefore we may set

$$N(k, s, n) = [s + N(k-1, s, n)]\left[N(k^{N(k-1, s, n)+s}, s, n-1) + 1\right].$$

The lemma is proved.

DEFINITION 4. An element b of a free associative ring $\mathfrak{A}$ on a set R of generators will be called a *J-polynomial* if it may be represented in the form of a polynomial in the elements of the set R relative to addition and the Jordan multiplications a^2 and aba.

For example, $\{abc\} = abc + cba$ and $a \circ b = ab + ba$ and all powers a^t are J-polynomials.

DEFINITION 5. An associative word α on elements of the set R will be called *distinguished* if there exists a homogeneous J-polynomial b_α whose lexicographically leading word is the word α, appearing in the polynomial b_α with *coefficient* 1.

LEMMA 4. *Each T-word α is distinguished (relative to the set R).*

PROOF. If the T-length of the word equals 1, i.e. $\alpha = a_k^t a_{i_1} \cdots a_{i_m}$ ($i_r \neq k$ for $r = 1,\ldots,m$, $m > 1$), then

$$b_\alpha = \left(\left(\left(a_k^t \circ a_{i_1}\right) \circ a_{i_2}\right) \cdots \right) \circ a_{i_m}.$$

Let the assertion of the lemma be proven for all T-words of T-length less than the T-length of the word α, which is larger than 1. Then

$$\alpha = \beta a_k^t a_{i_1} a_{i_2} \cdots a_{i_m} \qquad (i_r \neq k \text{ for } r = 1,\ldots,m),$$

where β is a T-word satisfying the condition of the inductive hypothesis. Let b_β be the J-polynomial corresponding to the word β. Then a simple calculation shows that we may take as b_α the J-polynomial

$$\left(\left(\left\{b_\beta a_k^t a_{i_1}\right\} \circ a_{i_2}\right) \cdots \right) \circ a_{i_m}.$$

The lemma is proved.

Consider an arbitrary ring K, which generally speaking will be nonassociative. Let Γ be some subsemigroup of the additive group of the ring K, and let its elements satisfy some homogeneous *polynomial* identity

$$f\left(\gamma_1^{p_1}, \gamma_2^{p_2}, \ldots, \gamma_k^{p_k}\right) = 0 \quad \text{for all } \gamma_1 \in \Gamma.$$

Here by $f(x_1^{p_1}, \ldots, x_k^{p_k})$ we denote a (nonassociative) homogeneous polynomial in the variables x_i ($i = 1,\ldots,k$), in each monomial of which x_i appears p_i times, having as coefficients elements of some arbitrary associative ring Σ of operators on k.

DEFINITION 6. *The multilinear polynomial*

$$\bar{f}(x_{11},\ldots,x_{1p_1};\; x_{21},\ldots,x_{2p_2};\ldots;x_{k1},\ldots,x_{kp_k})$$

corresponding to the polynomial $f(x_1^{p_1}, x_2^{p_2},\ldots,x_k^{p_k})$ is the polynomial obtained from f by replacing each variable x_i by one of the variables x_{is}, so that in each monomial there appears one and only one x_{is}, and afterwards summing up over all permutations of the symbols $x_{i1}, x_{i2},\ldots,x_{ik}$ for all $i = 1,\ldots,k$.

For example, if $f(x_1^3, x_2^2) = \{(x_1x_2)x_1\}(x_1x_2)$, then

$$\bar{f}(x_{11}, x_{12}, x_{13}, x_{21}, x_{22}) = [(x_{11}x_{21})x_{12}](x_{13}x_{22}) + [(x_{11}x_{22})x_{12}](x_{13}x_{21})$$
$$+ [(x_{11}x_{21})x_{13}](x_{12}x_{22}) + [(x_{11}x_{22})x_{13}](x_{12}x_{21}) + [(x_{12}x_{21})x_{11}](x_{13}x_{22})$$
$$+ [(x_{12}x_{22})x_{11}](x_{13}x_{21}) + [(x_{12}x_{21})x_{13}](x_{11}x_{22}) + [(x_{12}x_{22})x_{13}](x_{11}x_{21})$$
$$+ [(x_{13}x_{21})x_{11}](x_{12}x_{22}) + [(x_{13}x_{22})x_{11}](x_{12}x_{21}) + [(x_{13}x_{21})x_{12}](x_{11}x_{22})$$
$$+ [(x_{13}x_{22})x_{12}](x_{11}x_{21}).$$

LEMMA 5. *Any elements* γ_{ij} $(i = 1,\ldots,k; j = 1,\ldots,p_i)$ *of the semigroup* Γ *in the ring K satisfy the identity*

$$\bar{f}(\gamma_{11},\ldots,\gamma_{1p_1};\; \gamma_{21},\ldots,\gamma_{2p_2};\ldots;\gamma_{k1},\ldots,\gamma_{kp_k}) = 0.$$

PROOF. Let $p_1 = p_2 = \cdots = p_{s-1} = 1$ and $p_s > 1$. From properties of the semigroup Γ it follows that the polynomial

$$f\left(x_1, x_2,\ldots,x_{s-1},(x_{s1} + x_{s2} + \cdots + x_{sp_s})^{p_s}, x_{s+1}^{p_{s+1}},\ldots,x_k^{p_k}\right)$$

$$- \sum_{q=1}^{p_s} f\left(x_1,\ldots,x_{s-1},\left[\left(\sum_{j=1}^{p_s} x_{sj}\right) - x_{sq}\right]^{p_s}, x_{s+1}^{p_{s+1}},\ldots,x_k^{p_k}\right)$$

$$+ \sum_{p_s \geq q_1 \geq q_2 \geq 1} f\left(x_1,\ldots,x_{s-1},\left[\left(\sum_{j=1}^{p_s} x_{sj}\right) - x_{sq_1} - x_{sq_2}\right]^{p_s}, x_{s+1}^{p_s},\ldots,x_k^{p_k}\right)$$

$$\cdots + (-1)^{p_s-1} \sum_{t=1}^{p_s} f\left(x_1, x_2,\ldots,x_{s-1}, x_{st}^{p_s}, x_{s+1}^{p_{s+1}},\ldots,x_k^{p_k}\right)$$

specializes to zero when the variables are replaced by arbitrary elements of the semigroup Γ.

A simple calculation, carried out for each monomial of the polynomial f, shows however that this polynomial is linear in each variable x_{si} $(i = 1,\ldots,p_s)$ and is obtained from the polynomial f by substituting for each of the variables x_s in each term one of the variables x_{si} so that in each of the monomials of the polynomial f each x_{si} appears once and only once, and afterwards summing over all permutations of the symbols $x_1,\ldots,x_{sp_s}$. Carrying out this construction for each s in succession from 1 to k, we obtain what is required. The lemma is proved.

REMARK. This simple assertion is repeatedly met in a weaker formulation in algebraic works; however, it is usually proven for algebras (see, for example, [2]) with some restrictions on the base field.

§3. Special Jordan rings and algebras

We consider *special Jordan rings J*, i.e. rings imbedded in some associative ring $A_0(J)$, such that the set of elements corresponding to the elements of J form a Jordan ring (isomorphic to the ring J) relative to addition and the Jordan multiplications a^2 and aba. If for some element c of the ring J there exists a natural number $n(c)$ such that $c^{n(c)-1} \neq 0$ but $c^{n(c)} = 0$, we say as usual that the element c is *nilpotent of index $n(c)$*.

DEFINITION 7. If all elements of the ring J are nilpotent, and the totality of their indexes is bounded, then the ring J is called *nil of bounded index*.

DEFINITION 8. An arbitrary ring S is called *nilpotent* if there exists a natural number $N(S)$ such that the product of any $N(S)$ elements of the ring S under any method of arranging parentheses equals zero.

THEOREM 1. *Each special Jordan algebra, nil of bounded index with a finite number of generators, is nilpotent.*

DEFINITION 9. The *enveloping associative ring $A(J)$* of a special Jordan ring J is the intersection of all subrings of the ring $A_0(J)$ which contain J.

It is easily seen that the enveloping ring $A(J)$ is the subring generated in $A_0(J)$ by any set of generators for J.

The validity of Theorem 1 will follow from Theorem 2, which, broadly speaking, generalizes the theorem of Levitzki [7].

THEOREM 2. *The enveloping associative algebra $A(J)$ of a special Jordan ring J, nil of bounded index with a finite number of generators, is nilpotent.*(^{3})

PROOF. Let the ring J have a set $R = \{a_1, \ldots, a_k\}$ as a set of generators. We will take this same set R as set of generators for the ring $A(J)$. The proof of Theorem 2 will be carried out by induction, assuming it correct in case the number of generators of the ring J equals $k - 1$.

We consider an arbitrary R-word α of length $m[N(M, n, n) + 2]$, where m is the maximal length of nonzero a_k-indecomposable words (using the induction hypothesis), M is the number of such words, and n the bound on the index of the elements of J. Then in the word α there is a subword β which is a T-word and has T-length equal to $N(M, n, n)$.

By Lemma 3, in the word β either there are n consecutive equal subwords, or there is an n-split subword γ. We consider both these possibilities in succession.

1. $\beta = \beta_1 \gamma^n \beta_2$. By Lemma 4 the word γ is distinguished. Therefore γ is a leading term of a J-polynomial b_γ. Since $b_\gamma^n = 0$, β is expressed in the form of a

---

(^{3}) *Translator's note.* A more elegant way to say this is that the special universal envelope of J is nilpotent.

linear combination with integral coefficients of words of the same R-length, but (lexicographically) smaller than the word β. Consequently, α can be expressed in an analogous form.

2. $\beta = \beta_1 \gamma_1 \gamma_2 \cdots \gamma_n \beta_2$. The elements of the ring J form a subgroup of the additive group of the ring $A(J)$. By Lemma 5, from the relation $x_1^n = 0$ satisfied in $A(J)$ by all elements of J, it follows that the relation $\Sigma_p \, x_{i_1} \cdots x_{i_n} = 0$ is satisfied, where the summation applies to all permutations $(i_1, \ldots, i_n)$ of the symbols $1, \ldots, n$.

By Lemma 4 the elements γ_i are distinguished, hence the elements γ_i are the leading terms of J-polynomials b_{γ_i}.

Using the definition of n-split and the property of the J-polynomial b_{γ_i}, we see that from the relation $\Sigma_p \, b_{\gamma_{i_1}} \cdots b_{\gamma_{i_n}} = 0$ there follows the possibility of expressing the element β in the form of a linear combination with integral coefficients of words of the same R-length but (lexicographically) smaller than the word β. Consequently α may be expressed in an analogous form.

Thus we arrive at the conclusion that either the word α equals zero, or the element α can be represented in the form of a linear combination with integral coefficients of words having the same R-length as α, but less than α. Since a decreasing sequence of words of equal length breaks off, $\alpha = 0$. Theorems 2 and 1 are proven.

Not to change notation, we will assume that J is a special algebraic Jordan algebra over a field F and the degrees of the elements are bounded by a number n. In other words, each element of J is a root of some (associative) polynomial of degree n in one variable x with coefficients from the field F (for comparison see [1]).

Let $S_t(x_1, \ldots, x_t) = \Sigma \pm x_{i_1} \cdots x_{i_t}$ be the alternating sum of $t!$ terms obtained from the product $x_1 \cdots x_t$ by means of all possible permutations of the factors; the sign of each term depends on the evenness $(+)$ or oddness $(-)$ of the corresponding permutation.[4]

LEMMA 6. *The standard identity*

$$S_{2n-1}\left(a, a^2, \ldots, a^n, b_1, b_2, \ldots, b_{n-1}\right) = 0$$

holds for all elements a of the algebra J and $b_1, \ldots, b_{n-1}$ of any enveloping associative algebra $A(J)$.

PROOF. It is easily seen that each alternating sum S_t of the indicated form vanishes if any two of this arguments are equal.

On the other hand, by assumption for any element $a \in J$ there exist elements $\delta_i(a) \in F$ such that

$$a^n = \delta_1(a)a^{n-1} + \delta_2(a)a^{n-2} + \cdots.$$

---

[4] *Translator's note.* The author uses the notation P_t. In the English literature it is usually denoted S_t and called the *standard identity* of degree t.

Substituting this expression for a^n in the indicated relation concludes the proof of Lemma 6.

The proven relation is not trivial, i.e. it is not satisfied in every associative algebra. Indeed, no other term will be similar to the term $ab_1a^2b_2 \cdots a^{n-1}b_{n-1}a^n$, for example.

THEOREM 3. *Each enveloping algebra $A(J)$ of a special algebraic Jordan algebra J of bounded degree over a field F is locally finite, i.e. each finite subset of its elements generates a subalgebra of finite rank.*[5]

PROOF. Each subalgebra $A_Q(J)$ of the algebra $A(J)$ having a finite number of generators lies in some subalgebra $A_R(J)$ whose set of generators $R = \{a_1, \ldots, a_k\}$ consists of elements of the algebra J. We shall prove that the rank of $A_R(J)$ is finite, using induction on k. Let the subalgebras generated in $A(J)$ by $k - 1$ elements of J have finite rank. Then there exists a natural number such that each word of length $\geqslant m$ generated by the elements of the set $R' = R \backslash a_k$ may be represented in the form of a linear combination of words of smaller length. Consider an R-word α of length $(m + n)[N(M, n, \frac{1}{2}(n^2 + 3n - 2)) + 1]$, where M is the number of different a_k-indecomposable words not representable in the form of linear combinations of R-words having smaller R-length, and n the bound on the degrees of elements of J.

Theorem 3 will be proven if we show that the word α may be represented in the form of a linear combination of words having smaller R-length.

If $\alpha = \alpha'\beta\alpha''$, where α' is an R'-word, β a T-word, and $\alpha'' = a_k a_k \cdots a_k$, then we may assume that the R-lengths of the words α' and α'' are less, respectively, than the numbers m and n, since otherwise all would be proven. Then the R-length of the word β is larger than $(m + n)N(M, n, \frac{1}{2}(n^2 + 3n - 2))$. We may assume that each of the a_k-indecomposable words appearing in the word β may not be represented in the form of linear combinations of R-words of smaller R-length. Each such word has R-length less than $m + n$. Therefore, the T-length of the word β is larger than the number $N(M, n, \frac{1}{2}(n^2 + 3n - 2))$. On the ground of Lemma 3 we may assert that in the word β either there appear n consecutive equal T-subwords, or there appears an $\frac{1}{2}(n^2 + 3n - 2)$-split T-word. We consider both possibilities separately.

1. $\beta = \beta_1\gamma_1''\beta_2$. Using the algebraicity of the J-polynomial b_{γ_1}, defined analogously to that appearing in the proof of Theorem 2, we obtain an expression for the word β, and hence the word α, in the form of a linear combination of smaller words (in the sense of the lexiographic ordering) and words having smaller R-length.

---

[5] *Translator's note.* Again it might be more elegant to say the special universal envelope of J is locally finite. Further, the argument in Theorems 3 and 4 holds for algebras over an arbitrary ring of scalars Σ; we need only define *algebraic over* Σ to mean that some power a^n is a Σ-linear combination of lower powers, *locally finite* to mean that each finite subset generates a subalgebra which is finitely spanned over Σ, and to observe that if $A_R(J)$ is finitely spanned over Σ then any subalgebra $A_Q(J)$ is algebraic of bounded degree over Σ.

2. $\beta = \beta_1 \gamma_1 \gamma_2 \cdots \gamma_n, \beta_2$ for $n' = \frac{1}{2}(n^2 + 3n - 2)$. On the basis of Lemmas 6 and 5, the J-polynomials $b_{\gamma_{i_s}}$ $(s = 1, \ldots, n')$ satisfy a nonlinear identity of degree $n' = 1 + 2 + \cdots + n + (n - 1)$. It follows from this that the product $b_{\gamma_1} \cdots b_{\gamma_{n'}}$ may be expressed in the form of a linear combination of products, obtained from the given one by means of permuting the factors. As in the proof of Theorem 2, we conclude the possibility of expressing the word β, and hence α, in the form of a linear combination of smaller words.

Repeating the reasoning applied to the lexicographically smaller words obtained, in the end we obtain an expression for the word α in the form of a linear combination of words having smaller R-length. The theorem is proved. An obvious corollary of Theorem 3 is the following proposition:

THEOREM 4. *Each special algebraic Jordan algebra of bounded degree over a field F is locally finite.*

REMARK 1. The question of the possibility of omitting in the formulation of Theorems 1 and 4 the restriction to speciality remains open, though from the word of the author [3] and Theorems 1 and 4 it follows that any two elements generate, respectively, a nilpotent ring or an algebra of finite rank in the nil ring J of bounded index or in the algebraic Jordan algebra of bounded degree.

REMARK 2. If we deal only with *linear* Jordan algebras and Jordan multiplication $a \circ b = ab + ba$ the restriction on the additive group that there be no 2-torsion or on the characteristic of the base field that it be $\neq 2$ is essential, as is shown by the example of a free Lie algebra with 2 generators over a field of characteristic 2, which is a special linear Jordan algebra in view of the Birkhoff-Witt theorem (see [5] or [8]) but has infinite dimension. Note, however, that this example is not a (quadratic) Jordan algebra since it is not closed under *aba*.

§4. Right alternative and alternative rings and algebras

As is well known, a ring S is called *right alternative* (correspondingly, *left alternative*) if for any two of its elements a and b the equality $(ab)b = a(bb)$ holds (correspondingly, $b(ba) = (bb)a$), and *right Moufang* if in addition for any three elements a, b, c the equation $\{(ab)c\}b = a\{(bc)b\}$ holds (correspondingly, *left Moufang* if $b\{c(ba)\} = \{b(cb)\}a$). When there is no 2-torsion, the concepts of right alternative rings and right Moufang rings coincide. Rings which are simultaneously right and left alternative (equivalently, right and left Moufang) are called *alternative*. It is well known [9] that relative to the operations of addition and Jordan multiplications a^2 and $(ab)a$ the elements of a right Moufang ring form a special Jordan ring. From Lemma 5 it follows that the well-known multilinear identities

$$(1) \qquad (ab)c + (ac)b = a(bc) + a(cb)$$

and

$$(2) \qquad \{(ab)c\}d + \{(ad)c\}b = a\{(bc)d\} + a\{(dc)b\}$$

hold.

Fixing some set R of generators of a right Moufang ring S, we can carry Definition 4 of a J-polynomial over to elements of a right Moufang ring. For example, $a \circ b = ab + ba$ and $\{abc\} = (ab)c + (cb)a$ are J-polynomials, but the element $a(ba)$ need not be.

DEFINITION 10. Let $a_{i_1} \cdots a_{i_s}$ be some associative word on the elements of the set R. Then we set

$$\langle a_{i_1} a_{i_2} \cdots a_{i_s} \rangle = \{ \cdots [(a_{i_1} a_{i_2}) a_{i_3}] a_{i_4} \cdots \} a_{i_s}.$$

The operation $\{ \}$ is extended to nonassociative words by ignoring the parentheses existing within them, and to linear combinations of these words. For example

$$\langle (ab)(cd) + m[n(pq)] \rangle = [(ab)c] + [(mn)p]q.$$

A bar over any subword whatever or element will denote that the given subword or element is to be considered as a particular generator and does not undergo change. For example,

$$\langle [(ab)(\overline{cd})](mn) \rangle = \{ [(ab)(cd)]m \} n = \langle [\overline{(ab)(cd)}](mn) \rangle$$

and

$$\langle [(ab)(cd)](\overline{mn}) \rangle = \{ [(ab)c]d \}(mn).$$

If in the free nonassociative ring with generating set R the element q lies in the ideal generated by the elements of the form $(ab)c - a(bc)$, then it is obvious that $\langle q \rangle = 0$.

LEMMA 7. *If m is some element of a right Moufang ring S and d is some J-polynomial, then $md = \langle \overline{m}d \rangle$.*[6]

PROOF. In view of the last remark, it is enough to prove the lemma under the assumption that d is a J-monomial, i.e. can be written in the form of a Jordan product of some factors from the set R.

If d has degree 1, i.e. is an element of R, there is nothing to prove. Let d have degree $n > 1$, and let the assertion already be proven for smaller degrees. Then d may be expressed as d_1^2, $d_1 \circ d_2$, $(d_1 d_2)d_1$, or $\{d_1 d_2 d_3\}$ where d_1, d_2, and d_3 are J-monomials meeting the conditions of the induction assumption. Then respectively

$$md = m(d_1 d_1) = (md_1)d_1 = \langle \overline{m}d_1 \rangle d_1 = \langle (\overline{m}d_1)d_1 \rangle = \langle \overline{m}(d_1 d_1) \rangle = \langle \overline{m}d \rangle,$$

$$md = m(d_1 d_2 + d_2 d_1) = (md_1)d_2 + (md_2)d_1 = \langle (\overline{m}d_1)d_2 \rangle + \langle (m\overline{d}_2)d_1 \rangle$$
$$= \langle \overline{m}(d_1 \circ d_2) \rangle = \langle \overline{m}d \rangle,$$

$$md = m\{(d_1 d_2)d_1\} = \{(md_1)d_2\}d_1 = \langle \{(\overline{m}d_1)d_2\}d_1 \rangle = \langle \overline{m}\{(d_1 d_2)d_1\} \rangle$$
$$= \langle \overline{m}d \rangle,$$

$$md = m[(d_1 d_2)d_3 + (d_3 d_2)d_1] = [(md_1)d_2]d_3 + [(md_3)d_2]d_1 = \langle \overline{m}\{d_1 d_2 d_3\} \rangle$$
$$= \langle \overline{m}d \rangle.$$

---

[6] *Translator's note.* In terms of right multiplication operators R_x $(R_x y = yx)$ this may be formulated as $R_{f(x_1,\ldots,x_n)} = f(R_{x_1},\ldots,R_{x_n})$ for any Jordan polynomial f.

This uses relations (1), (2), the inductive assumption, and the linearity of the operator $\langle\ \rangle$. The lemma is proved.

LEMMA 8. *Under the hypotheses of Lemma 7, $d = \langle d \rangle$.*

PROOF.([7]) Using the scheme of proof of Lemma 7, and guided by it, we get the series of equalities

$$d = d_1^2 = \langle \bar{d}_1 d_1 \rangle = \langle d_1 d_1 \rangle = \langle d \rangle,$$
$$d = d_1 \circ d_2 = d_1 d_2 + d_2 d_1 = \langle \bar{d}_1 d_2 \rangle + \langle \bar{d}_2 d_1 \rangle = \langle d_1 d_2 \rangle + \langle d_2 d_1 \rangle = \langle d \rangle,$$
$$d = (d_1 d_2)d_1 = \langle\langle \bar{d}_1 d_2 \rangle d_1 \rangle = \langle(\bar{d}_1 d_2)d_1 \rangle = \langle(d_1 d_2)d_1 \rangle = \langle d \rangle,$$
$$d = \{d_1 d_2 d_3\} = (d_1 d_2)d_3 + (d_3 d_2)d_1 = \langle(\bar{d}_1 d_2)d_3 \rangle + \langle(\bar{d}_3 d_2)d_1 \rangle$$
$$= \langle\{d_1 d_2 d_3\}\rangle = \langle d \rangle,$$

proving the lemma.

DEFINITION 11. A monomial q of the free nonassociative ring on generators $R = \{a_1, \ldots, a_k\}$ is called an r_1-*word* if $q = \langle q \rangle$. By induction, an r_i-*word* is an r_1-word on r_{i-1}-words. For example, the words

$$\{[(a_1 a_2)a_3]a_4\}a_5 \qquad \text{and} \qquad (\{(a_1 a_2)[(a_2 a_1)a_3]\}a_4)(a_1 a_3)$$

are, respectively, r_1- and r_2-words.

LEMMA 9. *Any element b of an alternative ring S may be represented in the form of a linear combination (with integral coefficients) of r_2-words on any set R of generators for the ring S.*

PROOF. It obviously suffices to prove the lemma under the hypotheses that b is a monomial. For monomials of degree $\leqslant 3$ the assertion is trivial. Let the lemma be proven for degree $< n$, and let the degree of the monomial b equal n. Using these hypotheses we have $b = b'b'' = \Sigma_i\, b_{2i}(c_{2i} d_{1i})$, where c_{ji}, b_{ji} and d_{ji} are r_j-words.

Now it suffices to prove the lemma for monomials of the form $b_{2i}(c_{2i} d_{1i})$ of degree n. If the degree of the monomial $c_{2i} d_{1i}$ is $\leqslant 2$, then the assertion is correct in a trivial manner. We make a second inductive assumption that the lemma is true for monomials of degree n if the degree of the right factor is less than m.

Now let the degree of the monmial $b_{2i}(c_{2i} d_{1i})$ equal n, and the degree of the monomial $c_{2i} d_{1i}$ equal m for $3 \leqslant m < n$.

First assume that the degree of the monomial d_{1i} is 1. We require two well-known identities:

$$(3) \qquad\qquad (ab)c + (ba)c = a(bc) + b(ac),$$

$$(4) \qquad\qquad (ab)c + (cb)a = a(bc) + c(ba),$$

([7]) *Translator's note.* The lemma also follows immediately from Lemma 7 by setting $m = 1$ in the ring obtained by externally adjoining a unit element.

satisfied by any elements a, b, c of an alternative ring. Identity (3) is analogous to (1) and is satisfied in any left alternative ring; identity (4) (flexibility) is an immediate consequence of (1) and (3). Using in succession (4) and (1), we get

$$b_{2i}(c_{2i}d_{1i}) = -d_{1i}(c_{2i}b_{2i}) + (b_{2i}c_{2i})d_{1i} + (d_{1i}c_{2i})b_{2i}$$
$$= d_{1i}(b_{2i}c_{2i}) - (d_{1i}b_{2i})c_{2i} + (b_{2i}c_{2i})d_{1i}.$$

The monomials $(d_{1i}b_{2i})c_{2i}$ and $(b_{2i}c_{2i})d_{1i}$ satisfy the conditions of the second inductive hypothesis.

Consider the monomial $d_{1i}(b_{2i}c_{2i})$. Its factor $b_{2i}c_{2i}$ has degree $n - 1$; therefore $b_{2i}c_{2i} = \sum_t l_{2t}p_{1t}$. Using in succession (3) and (1), we have

$$d_{1i}(b_{2i}c_{2i}) = \sum_t d_{1i}(l_{2t}p_{1t}) = \sum_t \left[-l_{2t}(d_{1t}p_{1t}) + (d_{1i}l_{2t})p_{1t} + (l_{2t}d_{1i})p_{1t} \right]$$
$$= \sum_t \left[l_{2t}(p_{1t}d_{1t}) - (l_{2t}p_{1t})d_{1i} + (d_{1i}l_{2t})p_{1t} \right].$$

Since the degree of the monimial d_{1i} equals 1, and l_{jt} and p_{jt} are r_j-words, it is clear that all the derived words fall under the conditions of the inductive hypotheses ($p_{1t}d_{1i}$ is an r_1-word, and $(d_{1i}l_{2t})p_{1t}$ is expressible in terms of r_2-words since the right product of an r_2-word by an r_1-word is an r_2-word by Definition 11).

Thus, the lemma is proven under the inductive hypotheses and the additional assumption about d_{1i}. This forms a basis for a third inductive hypothesis that the lemma holds under the second inductive hypothesis in case the degree of the monomial d_{1i} is less than k. Now let this degree equal k, $1 < k < m$.

Preserving the previous sense of the indices, we have

$$b_{2i}(c_{2i}d_{1i}) = b_{2i}\left[c_{2i}(d_{1i}'a_s) \right],$$

where the degree of the monomial d_{1i}' equals $k - 1$ and $a_s \in R$. Using (1) and (2) in succession, we get

$$b_{2i}\left[c_{2i}(d_{1i}'a_s) \right] = -b_{2i}\left[(d_{1i}'a_s)c_{2i} \right] + \omega_1 = b_{2i}\left[(c_{2i}a_s)d_{1i}' \right] + \omega_2 = \omega_3$$

where the ω_i are linear combinations of monomials satisfying the inductive hypotheses. This completes the proof of Lemma 9.

DEFINITION 12. A ring S with a set R of generators will be called *right nilpotent relative to R* if there exists a natural number m such that $\langle d \rangle = 0$ for all R-monomials d of degree $\geq m$.

In view of the fact that each right Moufang ring S is power associative,[8] Definition 7 of nil ring carries over without change to right Moufang rings.

THEOREM 5. *Each right Moufang nil ring S of bounded index is locally right nilpotent relative to any set of generators.*

---

[8] I. M. Miheev [10] has given an example of a right alternative algebra of characteristic 2 which is not power associative.

Theorem 5 follows from the following assertion.

THEOREM 6. *If all J-polynomials of a right Moufang ring S on a set of generators $R = \{a_1, \ldots, a_k\}$ are nilpotent and the totality of their indexes is bounded, then the ring S is right nilpotent relative to the set R.*

PROOF. Let the number n be the bound on the indexes of the J-polynomials of the ring S. In the free associative ring A on the set of generators R consider the ideal I_1 generated by the nth powers of all J-polynomials. From Theorem 2 it follows that any monomial q of the ring A having degree $\geqslant m[N(M, n, n) + 2]$ belongs to I_1, i.e. $q = \Sigma_r l_r c_r j_r^n d_r$, where the l_r are integers, the c_r and d_r are monomials (which may be absent) and the j_r are J-polynomials.

Since the above equality is satisfied in the free associative ring, in any (nonassociative) ring on the same set of generators the equality $\langle q \rangle = \Sigma_r l_r \langle c_r j_r^n d_r \rangle$ will hold.

Using Lemma 7, and possibly Lemma 8, we obtain that in the ring S

$$\langle q \rangle = \sum_r l_r \langle \langle c_r j_r^n \rangle d_r \rangle = \sum_r l_r \langle \langle c_r \rangle \bar{j}_r^n d_r \rangle = 0$$

since each power of a J-polynomial is a J-polynomial.

Theorems 5 and 6 are proven.

THEOREM 7. *An alternative nil ring of bounded index is locally nilpotent.*

PROOF. By Theorem 6 each finite set $R = \{a_1, \ldots, a_k\}$ in the ring S generates a subring which is right nilpotent relative to R. The finite set R_1 of nonzero r_1-words on R generates a subring S_1 which is right nilpotent relative to R_1.

From the proof of Lemma 9 it follows that any R-monomial q in the ring S may be represented in the form of a linear combination with integer coefficients of r_2-words of the same R-length. If q is an r_2-word of large enough R-length, then from the right nilpotence of the ring S_1 it follows that $q = 0$. The theorem is proved.

LEMMA 10. *In any algebraic right Moufang algebra S of bounded degree over a field F, generated by a finite set R of generators, there are only a finite number of linearly independent r_1-words (relative to R).*

PROOF. For each J-monomial j_s an equality

$$f_s(j_s) = j_s^n + \delta_{s1} j_s^{n-1} + \delta_{s2} j_s^{n-2} + \cdots = 0$$

holds, where $\delta_{si} \in F$.

We consider the free associative algebra A over the field F with set of generators R, and in it the ideal I_1 generated by all possible elements $f_s(j_s)$. From Theorem 3 it follows that the factor algebra $\bar{A} = A/I_1$ is locally finite.

We show that linear dependence in the factor algebra $\bar{A}$ of the images of some words q_i ($i = 1, \ldots, t$) implies linear dependence of the r_1-words $\langle q_i \rangle$ in the algebra S. The first linear dependence is equivalent to a relation

$$\sum_{i=1}^{t} \mu_i q_i + \sum_s \rho_s c_s f_s(j_s) d_s = 0$$

in the free associative algebra A for $\mu, \rho \in F$ and c_s, d_s some monomials.

As in the proof of Theorem 6, we obtain

$$\sum_{i=1}^{t} \mu_i \langle q_i \rangle + \sum_s \rho_s \langle \langle c_s \rangle \overline{f_s(j_s)} \, d_s \rangle = 0,$$

from which follows the equality $\sum_{1}^{t} \mu_i \langle q_i \rangle = 0$ in the algebra S. The lemma is proved.

THEOREM 8. *Each algebraic alternative algebra S of bounded degree over a field F is locally finite.*

This follows at once from Lemmas 9 and 10.

BIBLIOGRAPHY

1. A. G. Kuroš, *Ringtheoretische Probleme, die mit dem Burnsideschen Problem über periodische Gruppen in Zusammenhangstehen*, Izv. Akad. Nauk SSSR Ser. Mat. **5** (1941), 233–240. (Russian; German summary)

2. A. I. Mal'cev, *On algebras defined by identities*, Mat. Sb. **26 (68)** (1950), 19–33. (Russian)

3. A. I. Širšov, *On special J-rings*, Mat. Sb. **38 (80)** (1956), 149–166. (Russian)

4. A. A. Albert, *Power-associative rings*, Trans. Amer. Math. Soc. **64** (1948), 552–593.

5. Garrett Birkhoff, *Representability of Lie algebras and Lie groups by matrices*, Ann. of Math. (2) **38** (1937), 526–532.

6. Nathan Jacobson, *Structure theory for algebraic algebras of bounded degree*, Ann. of Math. (2) **46** (1945), 695–707.

7. J. Levitzki, *On a problem of A. Kurosch*, Bull. Amer. Math. Soc. **52** (1946), 1033–1035.

8. Ernst Witt, *Treue Darstellung Liescher Ringe*, J. Reine Angew. Math. **177** (1937), 152–160.

9. R. D. Schafer, *Representations of alternative algebras*, Trans. Amer. Math. Soc. **72** (1952), 1–17.

10.* I. M. Miheev, *Right nilpotence in right alternative rings*, Sibirsk. Mat. Ž. **17** (1976), 225–227; English transl. in Siberian Math. J. **17** (1976).

Translated by K. McCRIMMON

---

* Added by translator.

Amer. Math. Soc. Transl.
(2) Vol. **119**, 1983

On Rings With Polynomial Identities*(^{1})

A. I. ŠIRŠOV

§1. Introduction

This article continues the author's paper [2]. Rather than repeating a considerable number of definitions and explanations of notation (which would occupy an unwarranted amount of space), the author limits himself to frequent references.

The first part of the work (§2) is devoted to associative rings with polynomial identities. Here Theorem 1 is proven, establishing a property of local finiteness for such rings, and some immediate corollaries of this theorem are indicated (which follow almost trivially from it); in particular the theorem of Kaplansky [4] on local finiteness of algebraic PI-algebras is obtained for algebras over an arbitrary ring of scalars. In the following §3 some generalizations of Kaplansky's theorem are proven for the case of alternative rings (Theorem 5).

§2. Associative rings with polynomial identities

At first we consider associative algebras over a field satisfying some (or maybe, one) polynomial identity. Serving as examples of such algebras are the commutative algebras and the finite-rank algebras. These examples show us how wide this class of algebras is, and how important it is to obtain general theorems formulated for arbitrary associative PI-algebras.

It is well known (see, for example, [1] or [2]) that in each such algebra we may assume some multilinear identity is satisfied. It is obvious that such an identity may always be written in the form

$$(1) \qquad x_1 x_2 \cdots x_n = \sum_{(i_1, i_2, \ldots, i_n)} \alpha_{i_1, \ldots, i_n} x_{i_1} x_{i_2} \cdots x_{i_n},$$

where the sum on the right is taken over all permutations $(i_1, \ldots, i_n)$ of the symbols $1, \ldots, n$ different from the identity permutation, and all α belong to the base field.

1980 *Mathematics Subject Classification.* Primary 16A38; Secondary 17D05.

*Translation of Mat. Sb. **43(85)** (1957), 277–283; MR **20** #1698. See the footnote on the first page of the preceding paper.

(^{1}) *Translator's note.* In the Russian literature these are known as "rings with identical relations".

DEFINITION 1. If some word s on symbols $v_1,\ldots,v_r$ may be written as $s = v_{i_1}^{m_1} \cdots v_{i_k}^{m_k}$ where $v_{i_s} \neq v_{i_{s+1}}$, the natural number k is called the *height* of the word s relative to the set $\{v_i\}$.

It is obvious that the word $s = v_1 v_1 v_2 v_1 v_1 v_2 v_1 v_1 v_2$, for example, has height 6 relative to the set $\{v_1, v_2\}$ and height 1 relative to the word $v_1 v_1 v_2$.

DEFINITION 2. In an algebra A with a finite number of generators $a_1,\ldots,a_l$ let there exist a set of elements $t_1,\ldots,t_k$, homogeneous in each a_i, such that each associative word s on the generators a_i equals when specialized in the algebra A some linear combination of words on the elements t_j which have the same composition (relative to the set $\{a_i\}$) as the word s,([2]) and which have height relative to the set $\{t_j\}$ no larger than some given number q. In this case, we will say that the algebra A has *bounded height*. If in an algebra B each finite subset generates a subalgebra of bounded height, we say that the heights of the algebra B are *locally bounded*.

It is obvious that the heights of every commutative algebra are locally bounded.

THEOREM 1. *The heights of any associative algebra satisfying a polynomial identity of degree n are locally bounded (relative to the set of words of degree $< n$ on any given set of generators).*

PROOF. Let the subalgebra A be generated by elements $a_1,\ldots,a_k$. According to Lemma 3 of [2] there exists a natural number $N = N(n)$ $(= N(k, 2n, n)$, in the notation of [2]) such that in each word of length N on the generators a_i there is either an n-split subword ([2], Definition 3) or a subword of the form b^{2n}.

We first prove that if the length m of the word b is at least n, then some subword of the word b^{2n} itself is n-split or of the form c^{2n} for c of length less than n, and in this connection it is natural to suppose the word b cannot itself be represented as a further power $\bar{b}^t$ for $t > 1$. From the word b we get m distinct words $b = b_0, b_1,\ldots,b_{m-1}$ by means of cyclic permutations of the factors.([3]) Suppose $b_{i_0} > b_{i_1} > \cdots > b_{i_{m-1}}$ in the lexicographic ordering. Since b^2 contains each b_{i_l} as a subword, it is obvious that the word b^{2n} may be represented in the form $b^{2n} = c b'_{i_0} b'_{i_1} \cdots b'_{i_{n-1}}$, where the first part of each of the words b'_i coincides with the corresponding word b_i. It is clear, further, that the subword $b'_{i_0} b'_{i_1} \cdots b'_{i_{n-1}}$ is m-split, and consequently it also contains an n-split subword.

Each n-split word, in view of relation (1), is representable in the form of a linear combination of lexicographically smaller words of the same composition. From this it follows that each word of length N on the generators a_i equals some

---

([2]) *Translator's note.* This means that these t_j-words, when written out in terms of the a_i, have the same factors as s but in a perhaps different order. For example, $s_{i_1} \cdots s_{i_n}$ has the same composition as $s_1 \cdots s_n = s$ for any a_i-subwords $s_1 \cdots s_n$ and any permutation $(i_1,\ldots,i_n)$.

([3]) *Translator's note.* The words $b_i = \sigma^i(b)$ for σ the cyclic permutation are distinct, since $\sigma^i(b) = \sigma^j(b)$ for $m > i > j$ would imply $\sigma^k(b) = b$, where $k = i - j$ has $m > k > 0$; but $b_k = b$ implies $b = (a_{i_1} \cdots a_{i_k})(a_{i_1} \cdots a_{i_k}) \cdots (a_{i_1} \cdots a_{i_k}) = \bar{b}^t$ for $\bar{b} = a_{i_1} \cdots a_{i_k}$, $t = m/k > 1$, contrary to hypothesis.

linear combination of words having the same composition relative to the genera-tors but containing subwords of the form b^{2n}, where b has length $< n$.

This last remark permits us to assert that if the height of a word s is great enough and this word does not contain an n-split subword, then it has a subword s_1 of the form $s_1 = b^n b'$, where the lengths m and m' of the words b and b' are connected by the inequality $n > m \geqslant m'$, and the word b' does not coincide with an initial subword of the word b.[4] In view of the finite number of possibilities for the words b and b' it follows easily that there exists a natural number M large enough so that each word $\bar{s}$ of height M (relative to the set of words of length $< n$) is a linear combination of words of the same composition that are lexicographically not greater than $\bar{s}$ and contain (not necessarily consecutively) n equal subwords of the form $s_1 = b^n b'$.[5]

However, in each such word there exists an n-split subword of one of the following forms:

$$\alpha_0\left(b^n b' \alpha_1 b\right)\left(b^{n-1} b' \alpha_2 b^2\right)\left(b^{n-2} b' \alpha_3 b^3\right) \cdots \left(b b' \alpha_n\right),$$
$$\alpha_0 b^n\left(b' \alpha_1 b^{n-1}\right)\left(b b' \alpha_2 b^{n-2}\right)\left(b^2 b' \alpha_3 b^{n-3}\right) \cdots \left(b^{n-1} b' \alpha_n\right)$$

depending on which of the words b or b' is lexicographically larger. Hence, each word of height $\geqslant M$ can be represented in the form of a linear combination of words of lower height. The theorem is proved.

We pass on to the consideration of associative Σ-Algebras (associative rings with operators from a commutative ring of scalars Σ).

DEFINITION 3. A polynomial identify satisfied by an associative Σ-algebra will be called *admissible* if (after reduction of similar terms) at least one of the coefficients of the terms of highest degree is 1.

THEOREM 2. *If an associative Σ-algebra S satisfies an admissible polynomial identity of degree n, then S has locally bounded height (relative to the set of words of degree less than n on any fixed set of generators).*

PROOF. In the chosen monic term of degree n of the polynomial identity, let the variables $x_1, \ldots, x_k$ appear with degrees $n_1, \ldots, n_k$ ($\Sigma_1^k n_i = n$) respectively. Carrying out the linearization of this term ([2], Lemma 5) with respect to $x_1, x_2, \ldots, x_k$

---

[4] *Translator's note.* The expression for a word on the a_i as a word on the monomials of degree $< n$ need not be unique, so height is not well defined; by height $> k$, we mean there is no expression whatever for the word having height $\leqslant k$.

If s has height $> N + 2$ in this sense, it has length $> N + 2$. The initial segment of length N contains a subword b^n for b of length $m < n$, so $s = s_0 b^n s_0'$, where s_0' has height > 2. We can write $s_0' = b^k b' s_0''$ and $b = b' b''$ (where b'' is not empty, though perhaps b' is), where s_0'' does not begin with an initial segment of b'' and has height > 0 (i.e. is nonempty). Then $s = s_0 b^k b^n b' s_0'' = s_0 b^k b'(b'' b')^n s_0'' = s_0 b^k b' \bar{b}_n s_0''$, where $\bar{b} = b'' b'$ has the same length m as b and s_0'' does not begin with an initial segment of $\bar{b}$. Taking $\bar{b}'$ to be an initial segment of s_0'' of nonzero length $m' \leqslant m$, we have our subword $\bar{s}_1 = \bar{b}^n \bar{b}'$.

[5] *Translator's note.* If there are r such words $b^n b'$ we may take $M = rn(N + 3)$. If $\bar{s}$ has height M we write $s = s_1 \cdots s_{rn}$, where s_k has height $N + 3$, and therefore contains some $b^n b'$; since there are rn of these but at most r distinct ones, at least one must appear $\geqslant n$ times.

in succession, we see that all terms will vanish in which any single x_i has degree less than n_i (when $n_i = 1$ we consider the identity $\phi(x_i, x_i') = f(x_i + x_i') - f(x_i) = 0$, where f is the left part of the given identity). Since in this process similar terms cannot appear, in the multilinear identity thus obtained at least one term will again have coefficient 1. Making, if necessary, a permutation of the variables, we obtain a relation of the form (1), which we used in the proof of Theorem 1. The rest of the proof repeats the proof of Theorem 1.

We consider some corollaries of our assertions.

THEOREM 3. *If all products of generators containing less than n factors are nilpotent in an associative Σ-algebra A with admissible polynomial identity of degree n, then the algebra A is locally nilpotent.*

The proof of this theorem is obvious.

COROLLARY 1. *If all products of degree $\leq n$ on generators of an associative algebra A of rank n over a field are nilpotent, then A is nilpotent.*

This assertion results from the well-known fact that any algebra of rank n satisfies the standard polynomial identity of degree $n + 1$ (the alternating sum of $(n + 1)!$ different products of $n + 1$ different variables is being annihilated).

DEFINITION 4. An element a of an associative ring A is called *algebraic* over some subring Z_1 of the center Z of A if there exist elements $z_i \in Z_1$ and a natural number m such that $a^m = \sum_{i=1}^{m-1} z_i a^{m-i}$.

DEFINITION 5. An associative ring A will be called *finite over the ring* Z_1, where Z_1 is a subring of the center Z of the ring A, if there exist elements $b_1, \ldots, b_k$ such that for some natural number m each element $c \in A^m$ may be represented in the form $c = \sum_1^k z_i b_i$, where the z_i are elements of the subring Z_1.([6])

As in the case of an algebra of finite rank, it is obvious that a ring finite over a subring of its center satisfies an admissible polynomial identity. From Theorem 2 we get at once the following more interesting proposition.

THEOREM 4. *If, in an associative ring A with a finite number of generators and an admissible polynomial identity of degree n, all products of less than n generators are algebraic over some subring Z_1 of the center of A, then the ring A is finite over Z_1.*

In the particular case when Z_1 is the zero subring, Theorem 4 includes the theorem of Levitzki [6]; it also includes the more general theorem of Kaplansky [4] (it is enough to externally adjoin a unit).

§3. Alternative and special Jordan rings with polynomial identities([7])

Henceforth we will consider alternative rings which satisfy some polynomial identity not a consequence of associativity.

([6]) *Translator's note.* Note that if A is infinite over Z_1 and also finitely generated, then A itself is finitely spanned over Z_1 by the b_i and all monomials of degree $< m$.

([7]) *Translator's note.* The author originally considered only rings without 2-torsion in this section.

DEFINITION 6. A polynomial identity $f(x_1,\ldots,x_n) = 0$ satisfied by an alternative ring is called *essential* if one of the coefficients of a leading term of the element $\langle f \rangle$ of the free nonassociative ring with generators x_i (see [2], Definition 9) equals 1 (after reduction of similar terms).

DEFINITION 7. A polynomial identity $I = 0$ in a special Jordan ring is called *admissible* if the relation $F = 0$ is admissible, where F is the associative polynomial obtained by developing the Jordan polynomial I.

LEMMA 1. *If the alternative ring K satisfies some essential polynomial identity, then the associated special Jordan ring K^+ satisfies some admissible polynomial identity.*

PROOF. Let $f(x_1,\ldots,x_q) = 0$ be the corresponding polynomial identity satisfied in K. Substituting the monomial xy^i in place of x_i in the polynomial f, we obtain some essential identity $\phi(x, y) = 0$. If $\bar{\phi}(x, y)$ is the polynomial obtained from the polynomial ϕ by writing the generators in each monomial in reverse order ([3], §3), then the ring K satisfies the admissible essential identity $\psi(x, y) = \phi(x, y)\bar{\phi}(x, y) = 0$. However, the polynomial $\psi(x, y)$ is a J-polynomial since $\bar{\psi} = \psi$ (see [3] and [5]) (in this we use the associativity of any alternative ring with 2 generators). If $\psi(x, y) \equiv I(x, y)$, then the J-polynomial $I(x, y)$ identically specializes to zero in the ring K^+. The lemma is proved.

DEFINITION 8. The *center* of a nonassociative ring T is the totality of all elements $x \in T$ such that $xa = ax$ and $(xa)b = x(ab) = a(bx)$ for all elements $a, b \in T$.

It is easily verified that the center Z is a subring.

REMARK. Having in mind the examination of the center of an alternative ring, generally speaking,([8]) in Definition 8 we could restrict the requirement to $xa = ax$. We do not do this, not wishing to divert the reader from the main problem of this work by the details that would arise here.

Definitions 4 and 5 carry over to alternative rings.

THEOREM 5. *An alternative ring K with a finite number of generators and an essential polynomial identity is locally finite over any subring Z_1 of its center over which all J-monomials on r_2-words on the generators are algebraic.*

PROOF. Let λ be the highest of the elements of the set R of generators of the ring K. We consider an r_1-word w on the elements of the set R in which λ appears less than $m(\lambda)$ times in succession, where $m(\lambda)$ is the degree of the elements λ. If the associative word $\bar{w}$, obtained from the word w by omitting parentheses, has a λ-decomposition, we denote by $l_\lambda(w)$ the number of λ-indecomposable factors obtained in this decomposition.

(^{8}) *Translator's note.* For example, when the ring has no 3-torsion or when it has no nilpotent elements.

Let k different λ-indecomposable words appear in the word $\bar{w}$, and let $l_\lambda(w) > N(k, s, n)$ ([2], Lemma 3), where n is the degree of the polynomial identity satisfied in the ring K, and $s \geq 2n$ is an upper bound on the degrees of all J-monomials $b_{\bar{v}}$ on the generators corresponding to subwords $\bar{v}$, generated by the above k different λ-indecomposable words, for which $l_\lambda(v) < n$. From Lemma 3 of [2] and the proof of Theorem 1 it follows that in the word $\bar{w}$ there is a subword u which has either the form $u_1 \cdots u_n$ or the form $u = (u')^s$, where u' and the u_i are the words generated by λ-indecomposable words, $l_\lambda(u') < n$, and $u_1 \cdots u_n$ is an n-splitting of the word u.

In each of these cases we may express the word w in the form of a linear combination of r_1-words lexicographically smaller than w, and words (with coefficients from the ring Z_1) whose R-lengths are smaller than the R-length of w.

In view of the fully analogous reasoning, we consider below only the first case.

Let $\bar{w} = \alpha u \beta$ where $u = u_1 \cdots u_n$. The word u_i (up to a factor of the form 2^t) is the leading associative word of some J-monomial b_{u_i} ([2], Lemma 4). Therefore, the element $W = \langle \alpha b_{u_1} \cdots b_{u_n} \beta \rangle = \langle \alpha \bar{b}_{u_1} \cdots \bar{b}_{u_n} \beta \rangle$ (for notation see [2], §4) has as leading term the word w. According to the lemma, there is a J-polynomial $I(x_1, \ldots, x_n)$ specializing identically to zero in the ring K and having as its leading (associative) word the word $x_1 \cdots x_n$. Since $\langle \alpha I(b_{u_1}, \ldots, b_{u_n}) \beta \rangle = 0$ and $\langle \alpha \bar{I}(b_{u_1}, \ldots, b_{u_n}) \beta \rangle = \langle \alpha I(b_{u_1}, \ldots, b_{u_n}) \beta \rangle$, the word w equals a linear combination of (lexicographically) smaller r_1-words.

By means of the reasoning set forth above, we proceed by induction on the number of generators and verify, first of all, that each sufficiently long λ-indecomposable word is a linear combination of shorter λ-indecomposable words, which justifies the introduction of the number k, and, secondly, that each sufficiently long r_1-word equals a linear combination (with coefficients from the ring Z_1) of shorter r_1-words.

This assertion carries over at once to r_2-words which, in view of Lemma 9 of [2], concludes the proof of Theorem 5.(^{9})

For examples we indicate two corollaries of Theorem 5.

COROLLARY 2. *An alternative algebraic algebra with an essential polynomial identity is locally finite.*

The proof follows from the fairly obvious fact that the external adjunction of a unit does not violate algebraicity and preserves some polynomial identity (for example, $f(x_1, \ldots, x_n) \cdot \Sigma(-1)^i \langle x_{i_1}, \ldots, x_{i_n} \rangle = 0$, where $f(x_1, \ldots, x_n) = 0$ is the initial relation, and $i = (i_1, \ldots, i_n)$ runs through all permutations of the symbols $1, \ldots, n$; $(-1)^i = \pm 1$ depending on the evenness of the permutation i).

COROLLARY 3. *The enveloping associative algebra of a special algebraic Jordan algebra with a finite number of generators and a polynomial identity has finite rank.*

(^{9}) *Translator's note.* The argument at this point is rather condensed. A more detailed version of the argument may be found in [7].

The proof follows from the fact that for the finiteness of the number of linearly independent r_1-words it is sufficient in the proof of Theorem 5 that there exists a polynomial identity for all J-polynomials on the generators.

BIBLIOGRAPHY

1. A. I. Mal'cev, *On algebras defined by polynomial identities*, Mat. Sb. **26(68)** (1950), 19–33. (Russian)

2. A. I. Širšov, *On certain nonassociative nil rings and algebraic algebras*, Mat. Sb. **41(83)** (1957), 381–394; English transl. in this volume.

3. ______, *On special J-rings*, Mat. Sb. **38(80)** (1956), 149–166. (Russian)

4. Irving Kaplansky, *Topological representations of algebras*. II, Trans. Amer. Mat. Soc. **68** (1950), 62–75.

5. P. M. Cohn, *On homomorphic images of special Jordan algebras*, Canad. J. Math. **6** (1954), 253–264.

6. J. Levitzki, *On a problem of A. Kurosch*, Bull. Amer. Math. Soc. **52** (1946), 1033–1035.

7.* Kevin McCrimmon, *Alternative algebras satisfying polynomial identities*, J. Algebra **24** (1973), 283–292.

Translated by K. McCRIMMON

---

*Added by translator.

Hopi Basket Weaving

Platypus with deelybobbers
Dead air on the dubba dubba
Baby you're the sexiest thing on two wheels
One Is a wheel of fire
The other is a wheel of fortune
Barnyard Man, Barnyard Man
You've got a way with words.

PRADA
MILANO
DAL 1913